Praise for the revised edition of
The Emotional Lives of Animals

"A tour de force into the hearts and minds of animals. Through artfully weaving scientific findings with stories about individual animals experiencing anguish, grief, worry, titillation, trust, joy, love, and much, much more, Bekoff brings the emotional lives of animals to life. You cannot read this book and not be changed by it."

— **Jessica Pierce**, bioethicist and author of
Who's a Good Dog? and *Run, Spot, Run*

"I am so glad that Marc has updated his very important and popular book.... In *The Emotional Lives of Animals*, he has pulled together the growing body of scientific evidence that supports the existence of a variety of emotions in other animals, richly illustrated by his own careful observations and conclusions. He argues forcefully that the time has come for acceptance of this body of information across the board."

— **Jane Goodall**, from the foreword

"One of the greatest mistakes of twentieth-century science was denying nonhumans' feelings and thoughts. Marc Bekoff has reminded us, with sensitivity and rigor, that nonhumans have not only a mental life, but a rich, emotional mental life — one important to acknowledge but also delightful to behold. *The Emotional Lives of Animals* should be on the bookshelf of anyone who lives with or around animals — which is everyone."
— **Alexandra Horowitz**, author of *Inside of a Dog* and *The Year of the Puppy*

"*The Emotional Lives of Animals* was brave when it was first published in 2007 because Bekoff's views were controversial; and it was essential because of the wealth of information the book brought together. Since then Bekoff's views have largely carried the day, but this second edition is even more essential than the first because of the way that it documents and makes accessible the explosion of knowledge that has helped to bring about this change."

— **Dale Jamieson**, director of the Center for Environmental and Animal Protection; affiliated professor of law, medical ethics, and bioethics; and professor emeritus of environmental studies and philosophy; New York University

"In this highly revised edition of his precedent-setting book, award-winning researcher Marc Bekoff offers a thorough review of what we know about the emotional lives of a dazzling array of nonhumans and why we must use what we know on their behalf. The time is now for a breakthrough; we must stop relying on timeworn, misleading claims that we don't really know what other animals are feeling and what they want and need from us. In fact, we've known for decades that we are failing countless animals by ignoring who they are and what they're asking of us — namely, to treat them with dignity and compassion."
— **Sy Montgomery**, award-winning author of *The Soul of an Octopus, Of Time and Turtles*, and many other books

Praise for the first edition of
The Emotional Lives of Animals

"Demonstrating the far-reaching implications for readers' relationships with any number of living beings, Bekoff's book is profound, thought-provoking and even touching."
— ***Publishers Weekly*** (starred review)

"A readable book equally charming and challenging."
— ***Booklist***

"Marc Bekoff ably presents the richness and variety of the emotions in nonhuman animals — and doesn't hesitate to draw the ethical conclusions implicit in his findings. I hope this book will be widely read by those who care about animals — and even more widely by those who don't."
— **Peter Singer**, professor of bioethics, Princeton University

"Move over, Darwin. And prepare to be moved. In *The Emotional Lives of Animals*, world-class scientist Marc Bekoff argues forcefully that our emotions are the gifts of our animal ancestors. Bekoff's new book itself is a gift that invites us to explore and appreciate the passionate lives of animals."
— **Marty Becker**, resident veterinarian on *Good Morning America* and author of *The Healing Power of Pets*

"A thought-provoking, compassionate, and scholarly work from one of the world's most eminent behavioral scientists."

— **Dr. Ian Dunbar**, founder of the
Association of Pet Dog Trainers and author of
Before & After Getting Your Puppy and *Barking Up the Right Tree*

"Marc Bekoff is one of those rare scientists who can talk real sense about animals because he is aware of being an animal himself. Read this wonderful book."

— **Mary Midgley**, author of
Animals and Why They Matter and *The Ethical Primate*

"A glorious, moving, important book to enjoy and share."

— **Ingrid Newkirk**, cofounder and president of
People for the Ethical Treatment of Animals (PETA)

"I firmly believe that the more we care for the happiness of others, the greater our own sense of well-being becomes. Therefore, I welcome Marc Bekoff's book *The Emotional Lives of Animals*."

— **His Holiness the Dalai Lama**

"A purposeful combination of science, anecdotal evidence, and memoir, *The Emotional Lives of Animals* is a work of rare beauty and insight by an author who has often found himself at odds with his professional colleagues. Fortunately, Bekoff chose to follow his own path and, in doing so, invites others to join him on a journey of humane stewardship."

— *Best Friends*

"*The Emotional Lives of Animals* is not a tearjerker but a gritty portrait of individual and societal values, painted with a heady blend of soul and science. But Bekoff's refreshing candor and solid credibility makes it a must-read for all."

— *City Dog Magazine*

"This thought-provoking book could very likely change your life."

— *Animals' Voice*

"A passionate, thoughtful book."

— *BBC Wildlife*

"This is a book that will empower and give hope; it guides the reader not into a pit of despair over the cruelties humans inflict on other animals but into a light, helping us seek the way forward."
— **Captive Animal Protective Society**, UK

"[Bekoff] presents both touching anecdotes and scientific evidence to support his case that animals do indeed have feelings — rich emotional lives, in fact."
— *Dog Fancy*

"Anyone interested in animal emotion will want a copy of this book. Bekoff speaks with the unique authority of an expert who is experiencing the success of a revolution in science and ethics that he helped make and that will endure (if we endure) as one of the signal achievements — along with the expansion of human rights and environmentalism — of the late twentieth century."
— *The BARk*

"In clear and convincing language, Marc Bekoff provides rational defense for what many of us already sense — that animals can feel sorrow, joy, anger, and pleasure much as we humans do. Bekoff proves that this idea is not only compatible with the fact of evolution, but it is required by it. Once it considers the argument of this finely argued book, science will never be the same."

— **David Rothenberg**, professor of philosophy,
New Jersey Institute of Technology,
and author of *Why Birds Sing* and *Sudden Music*

The Emotional Lives of Animals

Also by Marc Bekoff

The Animal Manifesto: Six Reasons for Expanding Our Compassion Footprint

Animal Passions and Beastly Virtues: Reflections on Redecorating Nature

The Animals' Agenda: Freedom, Compassion, and Coexistence in the Human Age (with Jessica Pierce)

Animals at Play: Rules of the Game (a children's book)

Animals Matter: A Biologist Explains Why We Should Treat Animals with Compassion and Respect

Canine Confidential: Why Dogs Do What They Do

Dogs Demystified: An A-to-Z Guide to All Things Canine

A Dog's World: Imagining the Lives of Dogs in a World without Humans (with Jessica Pierce)

Minding Animals: Awareness, Emotions, and Heart

Nature's Life Lessons: Everyday Truths from Nature (with Jim Carrier)

Rewilding Our Hearts: Building Pathways of Compassion and Coexistence

Species of Mind: The Philosophy and Biology of Cognitive Ethology (with Colin Allen)

Strolling with Our Kin (a children's book)

The Ten Trusts: What We Must Do to Care for the Animals We Love (with Jane Goodall)

Unleashing Your Dog: A Field Guide to Giving Your Canine Companion the Best Possible Life (with Jessica Pierce)

Wild Justice: The Moral Lives of Animals (with Jessica Pierce)

Why Dogs Hump and Bees Get Depressed: The Fascinating Science of Animal Intelligence, Emotions, Friendship, and Conservation

Edited by Marc Bekoff

Animal Play: Evolutionary, Comparative, and Ecological Perspectives (with John Byers)

The Cognitive Animal: Empirical and Theoretical Perspectives on Animal Cognition (with Colin Allen and Gordon Burghardt)

Coyotes: Biology, Behavior, and Management

Encyclopedia of Animal Behavior

Encyclopedia of Animal Rights and Animal Welfare

Encyclopedia of Human-Animal Relationships: A Global Exploration of Our Connections with Animals

Ignoring Nature No More: The Case for Compassionate Conservation

Jane Goodall at 90: Celebrating an Astonishing Lifetime of Science, Advocacy, Humanitarianism, Hope, and Peace (with Koen Margodt)

The Smile of a Dolphin: Remarkable Accounts of Animal Emotions

The Emotional Lives of Animals

A Leading Scientist Explores Animal
Joy, Sorrow, and Empathy —
and Why They Matter

Marc Bekoff

Foreword by Jane Goodall

Revised Edition

New World Library
Novato, California

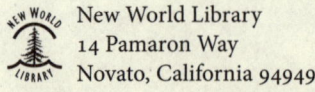 New World Library
14 Pamaron Way
Novato, California 94949

Copyright © 2007, 2024 by Marc Bekoff
Foreword copyright © 2007, 2024 by Jane Goodall

All rights reserved. This book may not be reproduced in whole or in part, stored in a retrieval system, or transmitted in any form or by any means — electronic, mechanical, or other — without written permission from the publisher, except by a reviewer, who may quote brief passages in a review.

Text design by Tona Pearce-Myers

Library of Congress Cataloging-in-Publication Data

Names: Bekoff, Marc, author.
Title: The emotional lives of animals : a leading scientist explores animal joy, sorrow, and empathy — and why they matter / Marc Bekoff ; foreword by Jane Goodall.
Description: Revised edition. | Novato, California : New World Library, [2024] | Includes bibliographical references and index. | Summary: "The author, a professor emeritus of ecology and evolutionary biology, demonstrates that animals have rich emotional lives. He presents recent animal behavioral research and narrates extraordinary stories of animal joy, empathy, grief, embarrassment, anger, and love"-- Provided by publisher.
Identifiers: LCCN 2023053725 (print) | LCCN 2023053726 (ebook) | ISBN 9781608689194 (paperback) | ISBN 9781608689200 (epub)
Subjects: LCSH: Emotions in animals.
Classification: LCC QL785.27 .B45 2024 (print) | LCC QL785.27 (ebook) | DDC 591.5--dc23/eng/20231213
LC record available at https://lccn.loc.gov/2023053725
LC ebook record available at https://lccn.loc.gov/2023053726

Originally published by New World Library in 2007
First revised edition printing, April 2024
ISBN 978-1-60868-919-4
Ebook ISBN 978-1-60868-920-0
Printed in Canada on 100% postconsumer-waste recycled paper

 New World Library is proud to be a Gold Certified Environmentally Responsible Publisher. Publisher certification awarded by Green Press Initiative.

10 9 8 7 6 5 4 3 2 1

For Jane Goodall on her ninetieth birthday and for all nonhuman animal beings and the wonderful people who work tirelessly to give them the best lives possible. They need all the help they can get.

Contents

Foreword by Jane Goodall..xiii
Introduction: The Gift of Animal Emotions.......................................1

Chapter 1: The Indisputable Case for Animal Emotions
  and Why They Matter..13
Chapter 2: Cognitive Ethology: Studying Animal Minds and Hearts.........41
Chapter 3: Beastly Passions: What Animals Feel............................55
Chapter 4: Wild Justice, Empathy, and Fair Play:
  Finding Honor among Beasts...93
Chapter 5: Addressing Uncertainty and Anthropomorphism
  in the Study of Animal Minds..117
Chapter 6: Ethical Choices: Why Animal Well-Being Matters..................133

Afterword: Compassion and Justice for All..................................183
Acknowledgments..191
Endnotes..193
Bibliography..224
Index..233
About the Author..253

Foreword

In 2007, I was very pleased to write a foreword for the first edition of this important book, for it dealt with a subject — emotion in nonhuman animals — that I have always maintained is crucial to a proper understanding of animals and their relationship to ourselves. Now, some seventeen years after the first edition of *The Emotional Lives of Animals* appeared, I'm pleased that Marc is updating what we know about the cognitive, emotional, and moral lives of a wide range of nonhuman animal beings.

We've come a long way in the past years. There is a great deal more awareness, and the amount of new information coming in, not only from scientific research, but that gained from the observations of the general public, is extraordinary. Many more people now understand that animals are way more intelligent than science once acknowledged, and that we humans are not, after all, the only sentient sapient beings on the planet. Understanding this makes it increasingly difficult to treat animals as mere "things." Instead we must respect them for who they are and think twice about interfering with their lives in ways that will cause them to feel fear and pain.

As examples, in 2012, the Cambridge Declaration on Consciousness was signed by a group of well-known scientists who agreed: "Convergent evidence indicates that nonhuman animals have the neuroanatomical, neurochemical, and neurophysiological substrates of conscious states along with the capacity to exhibit intentional behaviors. Consequently, the weight of evidence indicates that humans are not unique in possessing the neurological substrates that generate consciousness. Nonhuman animals, including all mammals and birds, and many other creatures, including octopuses, also possess these neurological substrates." They could also have included fishes and invertebrates, for whom evidence supporting sentience and consciousness is gradually mounting. Highly intelligent octopuses have "brains" very much unlike ours, and some researchers think they have brains in all eight legs that can act independently. It also looks like they can have nightmares and bad dreams.

In the past decade, several countries, cities, and towns have declared animals, including members of wild species, household companions, and farmed animals, to be sentient beings, and the field of compassionate conservation has emerged and is rapidly growing globally. In November 2022, the Jane Goodall Institute put forth a statement on keeping cetaceans in captivity, calling for "an immediate worldwide, permanent ban on capturing, keeping, and breeding cetaceans in captivity."

I've had a lifelong connection to many animals. Throughout my childhood I was fascinated with animals of all sorts — I watched them, learned from them, and loved them. At the age of ten, I developed a very special relationship with an extraordinarily intelligent mixed-breed dog, Rusty, who became my constant companion. There were other animals: three successive cats, two guinea pigs (whom I harnessed and took for walks), one golden hamster (who decided to live in the sofa), one canary and one budgerigar (both were allowed to fly freely in the house), and two rescued tortoises. With all, we shared our house and our hearts. And they taught me that animals, at least those with reasonably complex brains, have vivid and distinct personalities, minds capable of some kind of rational thought, and above all, feelings.

Then, in 1960, I had the extraordinary opportunity to learn about the chimpanzees of Gombe National Park in Tanzania. Knowing nothing of scientific method, I simply recorded everything I saw (writing in small notebooks with pen or pencil), as I had throughout my childhood. It was fortunate that I was patient, for during the first four months the chimps fled whenever they saw the strange white ape who had appeared so suddenly in their midst. The first individual to lose his fear I named David Greybeard. He was a strikingly handsome adult male, with large eyes set far apart. With his gentle but determined personality he was, as I ultimately discovered, a real leader — not because he was top-ranking or dominant, but because other chimps chose to follow him when he moved around the mountains. David's calm acceptance of my presence helped other members of his community to realize that I was not, after all, such a frightening creature. Then many of them became aggressive, treating me to the kind of intimidation displays normally directed at leopards or large snakes. But eventually they relaxed, and as I gradually gained their trust, they allowed me to move into their world — always on their terms. I got to know the various vivid personalities: David's close companion Goliath, who was, as I eventually realized, the alpha male; high-ranking, assertive Flo and her large family; timid Olly with her far-from-timid daughter, Gilka; irritable JB; Jomeo, the inadvertent clown — and all the rest.

After a year, Louis Leakey arranged for me to go to Cambridge University to work toward a PhD in ethology. There I was criticized for my lack of scientific method, for naming the chimpanzees rather than assigning each a number, for "giving" them personalities, and for maintaining they had minds and emotions. For these attributes, I was told sternly, were only present in the human animal. This was also the view of many philosophers and theologians. I was also told I should not have empathy with my "subjects," for scientists must be objective, and this was impossible if you were emotionally involved. This of course is not true.

When I submitted my first scientific paper for publication in *Nature*, the editor crossed out everywhere I had written "he" or "she" and substituted "it." Were they even to be deprived of their gender? I crossed out each *it* and underlined every *he* and *she*. And that was published.

How fortunate that Rusty had come into my life. For he, and my other animal companions, had taught me, clearly, that we humans are not the only beings with personalities, minds, and emotions. I simply did not accept that the behavior of other-than-human animals was for the most part merely a response to some environmental or social stimulus, that all their behavior was innately determined. That idea absolutely contradicted all I had learned during my years with Rusty and my new experiences with the chimpanzees.

Fortunately, I had a wise thesis supervisor, Professor Robert Hinde. He himself was known for his rigorous scientific mind and his intolerance of fuzzy thinking. Yet he had named all the rhesus monkeys he was studying and wrote of them, unashamedly, as he and she. It was Robert Hinde who taught me to express my commonsense but ethologically revolutionary ideas in a way that would protect me from too much hostile scientific criticism. For example, I could not say, "Fifi was happy," since I could not prove this, but I could say, "Fifi behaved in such a way that, had she been human, we would say she was happy." It is interesting that Robert's view changed after he spent two weeks at Gombe.

It is partly because I had had no university education that I was chosen to study the chimpanzees by my mentor, the late Professor Louis S. B. Leakey. It had been that very lack of qualifications, along with my passion for learning about animals in the wild, that had appealed to my mentor. He wanted an observer whose mind was unbiased by the reductionist thinking of ethologists at the time. Moreover, he felt that a woman might do better in the field, might be more patient.

During the late sixties, more and more biologists went into the field and started long-term studies on all manner of animal species: wolves, apes, monkeys, elephants, whales, dolphins, and so on. These studies, along with studies of pigs, rats, crows, and parrots in captivity, made it clear that animal behavior is far more complex than was originally admitted by Western science. There was increasingly compelling evidence that we are not alone in the universe, not the only creatures with minds capable of solving problems, capable of love and hate, joy and sorrow, fear and despair. Certainly we are not the only animals who experience pain and mental and physical suffering. In other words, there is no sharp line between the human animal and the rest of the animal kingdom. It is a blurred line, and becoming more so all the time.

Yet unfortunately, there are countless people among both the scientific and lay communities who still genuinely believe that animals are just objects, activated by instinctive responses to environmental stimuli. And only too often these people, consciously or unconsciously, reject our attempts to persuade them otherwise. After all, it is easier to do unpleasant things to unfeeling objects — to subject them to painful experiments in the laboratory, raise them in intensive factory farms, and hunt, trap, eat, and otherwise exploit them — than it is to do these things to sapient, sentient beings. Fear in a monkey, a dog, or a pig being is probably experienced in much the same way as it is in a human being. Young animals, human or otherwise, show such similar behavior when they are well fed and secure — frisking, gamboling, pirouetting, bouncing, somersaulting — that it is hard to believe they are not expressing very similar feelings. They are, in other words, full of joie de vivre — they are happy. Then again I have watched chimpanzee children, after the death of their mothers, show behavior similar to clinical depression in grieving human children — hunched posture, rocking, dull staring eyes, lack of interest in events around them. If human children can suffer from grief, so, too, can chimpanzee children. Sometimes, in this state of grieving, chimpanzee orphans — like Flint and Kristal at Gombe — may die.

Foreword

It is becoming increasingly obvious, and there is now excellent scientific backing to support this, that animals can be very beneficial for us. They can be very therapeutic, very healing. They play an important role in decreasing high blood pressure, reducing antisocial behavior in prisoners, and helping children with learning disabilities to read. Elderly people, living alone, can be saved from depression caused by loneliness, or feelings of uselessness, when they share their lives with a beloved cat or dog. This is not just because animals are soft, furry, and warm. It is because these animal healers seem to empathize with their humans, understand their needs — and love them. These animals, in other words, are a great deal more than objects whose behavior is triggered by stimulus and response. A mechanical stuffed toy animal, no matter how skillfully crafted, no matter how lifelike it appears, will never take the place of a living, feeling, and loving animal.

As more people understand the true nature of animals, the sooner we may succeed in changing the inappropriate ways in which so many millions of animals are treated. In fact, most people have no idea what goes on in medical research labs. They do not know — and do not want to know — about the cruelty of industrial farming, the billions of animals raised in stinking, unsanitary, and unbelievably cramped spaces in factory farms. Nor do they understand the cruelty involved in training animals to perform in circuses and other forms of entertainment. Unfortunately, as long as some scientists continue to uphold (at least in their professional lives) the mistaken view that other-than-human beings are mere things, this will be used to condone inhumane behavior of this sort.

That is why I am so glad that Marc has updated his very important and popular book. Undaunted by the sometimes vicious criticism from his peers that has been leveled at him throughout most of his professional life, he has continued to study and write about the personalities and emotions of animals. In *The Emotional Lives of Animals*, he has pulled together the growing body of scientific evidence that supports the existence of a variety of emotions in other animals, richly illustrated by his own careful observations and conclusions. He argues forcefully that the time has come for acceptance of this body of information across the board. He suggests, in fact, that it is a waste of time even to ask if chimpanzees, elephants, dogs, and so on experience happiness, jealousy, sadness, despair, and anger — that this is self-evident to everyone who has spent time with or shared their lives in a meaningful way with these animals. Instead of continuing to try to prove the obvious, surely the time has come to accept that animal beings, like human beings, express emotions, and to ask different questions — as he does in this book. Why did emotions evolve in the first place? What useful purpose do they serve? I think we should confront the skeptics by asking them to prove that animals do NOT have emotions.

The first edition of *The Emotional Lives of Animals* added a strong voice to the growing chorus of those who are trying to change attitudes toward the animal beings with whom we share this planet. The second edition adds to what we knew back then, likewise combining careful scientific methodology with intuition and common sense. This book will be a great tool for those who are struggling to improve the lives of animals in environments where, so often, there is an almost total lack of understanding. I only hope it will persuade many more people to reconsider the way they treat animals in the future.

— Jane Goodall, PhD, DBE, UN Messenger of Peace,
Founder of the Jane Goodall Institute

INTRODUCTION

The Gift of Animal Emotions

We like to see ourselves as special, but whatever the difference between humans and animals may be, it is unlikely to be found in the emotional domain.
— FRANS DE WAAL, "YOUR DOG FEELS AS GUILTY AS SHE LOOKS"

Since the first edition of this book was published in 2007, the field of cognitive ethology — the study of animal minds — has literally exploded, and now hardly a day goes by without there being some exciting new information about the cognitive, emotional, and moral lives of animals. In the first edition, I wrote, "Twenty years from now, our understandings and explanations will be richer, more accurate, and possibly different; this is true in any field." I had no idea I'd ever be writing a second edition, but seventeen years later, my prediction is accurate. This is absolutely true. And I'm very pleased and excited to have the opportunity to update this book and share this information with you.

I also wrote, "When it comes to the emotional lives of animals, science is just trying to catch up to what people experience every day." Science still is. I try to keep up with what's happening because I'm genuinely interested to learn and share as much as I can about the emotional lives of animals (in part through my *Psychology Today* blog, *Animal Emotions*). I often throw up my hands in joyful desperation because it's almost impossible to keep up, but this really is good news because so much of what comes my way confirms and expands on what I wrote nearly two decades ago.

As a scientist who's studied animal passions and beastly virtues for more than fifty years — really and truly — I consider myself very fortunate. I love what I do. I love learning about animals, and I love sharing with others what my colleagues and I discover. And I very much hope you enjoy this revised and expanded celebration of animal emotions and why they matter more than ever.

The Widespread and Growing Recognition of Sentience

The field of animal emotions — which is a specific area of focus within the larger scientific discipline of cognitive ethology — has changed a great deal over the past two decades. What I find especially interesting and gratifying is that one of the biggest changes is the widespread recognition of sentience in nonhuman animals. I like to think of sentience as being a "keystone concept" that draws together and supports individuals of many different species in the same manner that "keystone species" are essential to entire ecosystems, supporting the lives and well-being of individuals of many different species, who depend on them. We truly can speak about the "biodiversity of sentience."

New data from comparative research on an incredibly wide range of animals fully supports what I and many people have long proposed and believe about animal emotions. Of course, the central goals of this book remain the same: to explain the science of animal emotions, both what we know and how we know it, and to explore ways to use this knowledge to improve the lives of animals. However, one thing that is less necessary by the day is having to address skepticism about animal emotions themselves.

When I first began my studies some fifty years ago, researchers were almost all skeptics who spent their time wondering *if* dogs, cats, chimpanzees, and other animals — mostly mammals — felt anything. Since feelings don't fit under a microscope, these scientists usually didn't find any — and as I like to say, I'm glad I wasn't their dog! But thankfully, today, the burden of proof now falls most often on anyone who still argues that animals don't experience emotions. When writing about the inner lives of animals, my colleagues and I no longer have to put scare quotes around such words as *happy, sad, jealous, guilty, compassionate,* or *empathetic.* Scientific journals and the popular press regularly publish reports on joy in rats and grief in elephants, and no one blinks. And more and more scientific outlets are allowing researchers to refer to animals by name rather than by anonymous and distance-making numbers, a practice that Jane Goodall pioneered in her groundbreaking and ongoing research on wild chimpanzees.

I have always argued that stripping animals of emotions is antiscientific. While it's impossible to know precisely what any other animal is feeling — that is, to confirm any individual animal's private subjective experiences — it's also impossible to know any other *human's* private subjective experiences, but that doesn't lead us to deny common, shared, recognizable emotions among people. Primatologist Frans de Waal addressed this in his 2019 essay "Your Dog Feels as Guilty as She Looks": "For the longest time, science has depicted animals as stimulus-response machines while declaring their inner lives barren. This has helped us sustain our customary 'anthropodenial': the denial that we are animals."

I couldn't agree more. Based on the incredible amount of recent research on animal emotions, we can no longer deny that many other animals have rich and deep emotional lives, and further, we need to use what we know about their feelings on their behalf. That said, de Waal once proposed that talking as if we really knew what other animals feel required a leap of faith, and neuroscientist Jaak Panksepp had, in the past, seriously questioned what we really knew about animal emotions and what animals feel. But Panksepp once told de Waal, "There is solid evidence for animal feelings… [and] what's wrong with a few educated guesses?" Later, de Waal wrote, "I now believe he was right, and I will try to express his opinion and explain why he had to fight for it all his life."

For years I've argued that these "educated guesses" — actually, highly educated guesses — about animal emotions were not anthropomorphic projections but reflected as-yet-unproven facts that would one day be strongly supported by comparative scientific research. That day has come, and you'll find many examples in this updated book.

A watershed moment in the recognition of sentience was the 2012 Cambridge Declaration on Consciousness. This was drafted and signed by a wide range of eminent scientists who, in essence, stressed that it is about time we recognize that many nonhuman animals are conscious. Their declaration proposed that the real question is *why* consciousness has evolved, not *if* it has evolved.

Animal consciousness and sentience aren't science fiction. Solid science along with a touch of common sense clearly show that we surely are not the only animals who are aware of what is happening to us, around us, and to other individuals. And there is a growing list of countries and cities that have made their own formal declarations that animals are sentient beings. This list does not yet include the United States, but it does include Australia, Austria, Belgium, Bulgaria, Chile, Croatia, Cyprus, the Czech Republic, Denmark, Estonia, Finland, France, Germany, Greece, Hungary, Ireland, Italy, Latvia, Lithuania, Luxembourg, Malta, the Netherlands, New Zealand, Poland, Portugal, Romania, Slovakia, Slovenia, Spain, Sweden, Switzerland, and the United Kingdom.

In December 2021, according to one article: "Animals in Spain will no longer be considered as 'objects' by the law thanks to new legislation passed on Thursday by Spain's lower house, the Congress of Deputies. From now on, animals will be treated as 'sentient beings,' and as such will have a different legal status than an inanimate object. They will no longer be able to be seized, abandoned, mistreated, or separated from one of their owners in the case of a divorce or separation, without having their well-being and protection taken into account." In April 2022, Ecuador became the first country to give legal rights to individual wild animals, including the right to exist.

Sentience: The Ability to Feel and Why Feeling Matters

In a nutshell, here is what this book proposes and what it seeks to explain and demonstrate:

- Sentience simply means the capacity to feel.
- Recognizing and acknowledging sentience among nonhuman animals is no longer a radical view, as research has clearly shown that a wide variety of nonhuman animals experience a wide range of emotions, such as joy, sadness, grief, and so on.
- The important question is *why* sentience and emotions evolved among living beings, not *if* emotions evolved in some and not others.
- Skeptics no longer have a credible basis for blanket denials of sentience in nonhuman animals.
- We still have much to learn about sentience. Science is continually "surprising" us with evidence of sentience in unexpected species, such as fishes, amphibians, reptiles, insects, crustaceans, and possibly even plants.
- In other words, the more we study sentience, the more species we find that possess it, and we don't know all the forms it takes and ways it might be experienced.
- Once we acknowledge sentience in any species, it's important to

- use that knowledge to make decisions about the well-being of that species, including individuals.
- Talking about degrees of sentience makes little practical sense in regard to how we treat animals. An individual's pain is still pain. While there are differences among species, and differences among individuals, this does not mean that the suffering of certain animals is somehow less or doesn't count as much as the suffering of others.
- One of our biggest challenges today is aligning what we know about animal sentience and emotions with how we treat them. It is painful to acknowledge when we've caused harm, but the only solution is change. Society's current animal welfare standards are extremely outdated and inadequate.

Raise Your Paw (or Claw) If You Can Feel

In essence, what it all boils down to is that it's bad biology to argue against the existence of animal emotions. Scientific research in evolutionary biology, cognitive ethology, and social neuroscience supports the view that numerous and diverse animals have complex emotional lives. Emotions are an evolutionary gift; they have evolved as adaptations in numerous species, and they serve as a social glue to bond animals with one another. Emotions also catalyze and regulate a wide variety of social encounters among friends, lovers, and competitors, and they permit animals to protect themselves adaptively and flexibly using various behavior patterns in a wide variety of venues.

Charles Darwin's well-accepted ideas about evolutionary continuity — that differences among species are differences in degree rather than kind — also argue strongly for the presence of animal emotions, empathy, and moral behavior. In practice, continuity allows us to "connect the evolutionary dots" among different species to highlight similarities in evolved traits, including individual feelings and passions. What we have since learned about animal emotions and empathy fits well with what we know about the lifestyles of different species — how complex their social interactions and social networks are. Emotions, empathy, and knowing right from wrong are keys to survival, without which animals — both human and nonhuman — would perish. That's how important they are.

Further, all mammals (including humans) share neuroanatomical structures and neurochemical pathways that are important for feelings. For instance, it turns out that humpback whales, fin whales, killer whales, and sperm whales possess spindle cells in the same area of the brain as spindle cells in the human brain. Spindle cells were once thought to be unique to humans and other great apes, and they are believed to be important in processing emotions, playing a role in social organization, empathy, intuition, and rapid gut reactions. In fact, whales actually have more of them than humans do.

But even if many (if perhaps not all) animals share sentience, do all sentient animals feel the same things? We don't know yet, and we may never know for sure, but research is continually finding similarities. However, I think it's important to emphasize that, even as we recognize similarities among species, we shouldn't use ourselves as the template against which other animals are measured. The point isn't to compare animals to ourselves, but to understand animals for who they are. As the stories in this book show, mice are empathetic and fun-loving, and there are stories of pleasure-seeking iguanas, humorous horses, amorous whales, grieving otters, bereaved donkeys, pissed-off baboons, sentient fish, and elephants suffering from what mirrors post-traumatic stress disorder. Recent research has shown that elephants may be able to remember their relatives for around twelve years by smelling their dung. If that's so, there is no reason they couldn't also remember traumatic experiences for long periods of time.

Animals also clearly recognize feelings in other animals. That is, they possess what psychologists call emotional intelligence, or the ability to understand one's own emotions and those of others. There is an ever-growing number of stories about the love and devotion of longtime companions. To pick one among many stories, in 2023, two rescue dogs, a male named Heart and a female named Soul, were observed on the side of a highway near St. Louis, Missouri. Soul had been hit by a car, and Heart refused to leave her side. The attending veterinarians gave Heart the time he needed to say goodbye to his dear friend. They said, "This is true love." I urge you to have some tissues with you when you read this love story (see the endnotes). Later, I tell the story of Babyl, an elephant who couldn't have survived without the help of other caring and empathic group members.

While we might expect to find close, enduring, and endearing emotional relationships between members of the same species, improbable relationships often form between animals of wildly different species, even between animals who are normally predator and prey! Such was the case with Aochan, a rat snake, who befriended a dwarf hamster named Gohan, at Tokyo's Mutsugoro Okoku Zoo. These and other unlikely friendships are called "odd couples." Off the coast of Cornwall, England, bottlenose dolphins and harbor seals have been observed playing with one another. Unlikeliest of all? Perhaps Murphy, a quirky eagle living at the World Bird Sanctuary in Valley Park, Missouri, who formed a close relationship with a rock and incubated it as if he was to become a father.

Many species not only demonstrate emotional intelligence; they take actions that exhibit what we would consider our best human qualities: empathy, compassion, morality, and even altruism. Conducting research that confirms the presence of these attributes is difficult, but anecdotal evidence is mounting, such as in story collections of animals saving the lives of people and other animals. Books like *Daisy to the Rescue* by Jeff Campbell and *When Animals Rescue* by Belinda Recio recount dozens of real-life tales — such as of dolphins and whales saving people from sharks, lions rescuing a girl from kidnappers, a golden retriever fighting a cougar to save a boy, gorillas working together to dismantle poacher snares that were hurting their young and their elders, an elephant trying to rescue a baby rhino, and many more. We often speak of the need to save animals, but what does it say about animals when they recognize the need to save others?

And of course, humans and other animals also bond. I used to begin my lectures with the question: "Is there anyone in this audience who thinks that dogs don't have feelings — that they don't experience joy and sadness?" Even in scientific gatherings, few people ever raised their hands, though on occasion a hand or two would go up slowly, usually halfway, as the person glanced around to see if anyone was watching. Then I'd ask, "How many of you believe that dogs do have feelings?" Suddenly, almost every hand waved wildly and people smiled and nodded in vigorous agreement. As Jane Goodall describes with her dog Rusty, to live with a dog is to know firsthand that animals have feelings. It's a no-brainer. We map their feelings by observing their behavior, guided by the analogy of our own emotional templates, and we do it very reliably.

In essence, we feel a dog's emotions directly, and dogs feel ours. This nonverbal emotional exchange isn't imagined; it's real. As Frans de Waal notes, if we deny sentience just because animals can't speak, that would apply to "any organism that fails to talk," including infants and humans who can't talk. He says, "The importance we attach to language is just ridiculous. It has given us more than a century of agnosticism with regard to wordless pain and consciousness."

Nor is this bond exclusive to humans and dogs, or humans and domestic animals. This awareness and understanding exists between humans and all sentient animals, even with the wider natural world. For a dramatic account of this, read the memoir *Black Lion: Alive in the Wilderness* (see the bibliography), in which Zulu wilderness guide Sicelo Mbatha describes his encounters with wildlife and how he helps people see wildlife and nature with their hearts, not just their heads.

Nor, I might add, is a leap of imagination always out of place. Eight years ago, Ralph Nader asked me to write an endorsement for his 2016 novel *Animal Envy: A Fable*. I was very surprised, and my first thought was that this must be another Ralph Nader. This couldn't be *the* Ralph Nader, the environmental activist. I was wrong. I found *Animal Envy* to be a highly original, visionary, forward-looking book. Nader's story explores a very common wonder: What would animals tell us about themselves, and what would they say about us, if there really were a common language among all species? The story imagines that an app is developed that allows members of different species to talk among themselves and with us, and in a global assembly called "the Great Talkout," this is what happens. Based on research on the cognitive and emotional lives of animals, Nader weaves plausible stories of what animals might do and say and of what we might learn if all species had the ability to talk together — and if humans listened carefully.

Ever since researchers taught the gorilla Koko to use human sign language in the 1970s, we've tried to develop ways to talk with animals, and today, some scientists think AI might provide the communication breakthrough that makes Nader's novel no longer a "fable."

That said, the truth is, we and other animals already do communicate with one another, just indirectly — with our emotions. And animal behavior, at times, speaks volumes about the state of human-animal relationships; in and of itself, this behavior is often a plea for less violence and an appeal for compassionate coexistence. Animals want to live lives free from pain and suffering, and they have evolved all sorts of behavioral patterns and physical adaptations to help them along. We don't really need animals to speak our language to tell us this, since we already share the universal language of feelings.

We Need to Care for Animals and Our World

Recognizing that animals have emotions is important because their feelings matter. Their emotions matter to them, and they should matter to us. Animals are sentient beings who experience the ups and downs of daily life, and we must respect this when we interact with them. This includes not only the animals we live with, care for, and love — those we consider our "companions" — but also the billions of other domesticated animals who live on farms and die in slaughterhouses and provide us with food and clothing. And it includes wild animals who must continually struggle to share our ever-crowded world.

Needless to say, our relationships with other animals are complex, ambiguous, challenging, and frustrating, and we must continually reassess how we should interact with our nonhuman kin. Part of this reassessment involves asking difficult questions and making sure our actions match our understandings and beliefs. Thus, I often ask researchers who conduct invasive studies with animals, or I ask people who work on factory farms: "Would you do that to your dog?" Some are startled by this question, but it's a very important one to ask. If we wouldn't do something to our companion animals that we do daily to mice, rats, monkeys, pigs, cows, elephants, chimpanzees, or even noncompanion cats and dogs, we need to ask ourselves why.

I often write about the "empathy gap" that exists between us and other animals. Today, one of humanity's biggest challenges is overcoming this gap, for the sake of nonhuman animals, the world, and ourselves. This was brought home to me vividly a few years ago when, at a talk, a thirteen-year-old middle schooler, Monika, asked me if I thought we were a "failed mammal" because of how we treat other animals and other humans and all the environmental destruction we cause. I was shocked by her question, and it continues to haunt me. We often think we're better than other animals because we believe we're smarter, but that arrogance too often leads to disrespect for nonhuman animals and justifies the harm we cause. Our supposed superior intelligence has largely failed us when it comes to caring for our world. Or as one of my esteemed colleagues puts it, it has bitten us in the butt. This is something Justin Gregg muses about in his book *If Nietzsche Were a Narwhal: What Animal Intelligence Reveals about Human Stupidity*. In the long run, while human minds are fascinating and, in many ways, exceptional, they might be more curse than blessing when it comes to our ultimate survival and that of other species. My hope is that the jury is still out on whether that makes human intelligence an inferior evolutionary adaptation when compared to the kinds of intelligence we find in other species.

However, we have lots of work to do. For instance, consider Spain. While the country declared nonhuman animals to be "sentient beings" in 2021, and thus deserving of protection, bullfighting still exists in the country. In 2023, a Spanish business proposed creating an octopus factory farm that would kill a million octopuses annually by cruelly freezing them to death. The same year, at a Spanish hunting fair, fifty-four outfitters offered trophy hunting holiday packages for polar bears, lions, African elephants, giraffes, rhinos, and leopards. The majority of Spanish citizens oppose trophy hunting of internationally protected species, but these are examples of how difficult it is to translate declarations of sentience into societal change. Without changes in our behavior, what does a declaration of sentience really mean?

Or consider the blatant hypocrisy of US government agencies. The US Federal Animal Welfare Act redefined the word *animal* to exclude rats, mice, birds, fishes, and insects — all so that these species could continue to be abused and killed in scientific research without the protections given to other animals. Similarly, while the Bureau of Land Management and the US Fish and Wildlife Service claim their goal is to foster coexistence, they routinely slaughter countless animals they deem to be "pests." This is a perverted view of what *coexistence* means.

With Recognition Comes Dignity, Respect, and — Slowly — Protection

Nonetheless, declaring animals to be sentient beings helps move us in the right direction. Public acknowledgment of sentience encourages people to treat other animals with the respect, dignity, kindness, and compassion they fully deserve, and I'm pleased that more attention is being paid to using the word *dignity* to refer to nonhumans, both in the academic and popular press. And as more and more nonhumans are welcomed as full members into the sentience and consciousness club, that can inspire people to keep the pressure on so that declaring animals to be sentient beings really means something for the animals themselves. One example is the growing, global "Rights of Nature" movement. Advocates are seeking to replace humanity's exploitive relationship with nature — including both animals and ecosystems — with a legal partnership that respects the existence of rivers, forests, and all life. While recognizing animals as sentient beings may be slow to take hold, I agree with science writer Graham Lawton that "the world is still a better place because those rights exist."

Global conservation and rewilding efforts are also growing and changing our perspective on human-animal relationships in general, which is impacting the fields of anthrozoology, animal studies, conservation behavior, and conservation psychology. Researchers have learned that by protecting certain groups of important species, this helps protect entire ecosystems, and this can help meet planetary carbon-reduction goals.

Introduction: The Gift of Animal Emotions

Compassionate Conservation and the Importance of Individuals

Of course, protecting animals requires us to understand and respect their emotional lives and the importance of each and every individual. This has sparked the ever-growing field of compassionate conservation, which regards the life of every *individual* animal as a valued gift — and this paradigm shift can benefit all animals. Compassionate conservation asks us to consider what's best for each *individual* animal, whether in captivity, in our homes, or in the wild. All individuals have to be able to live the lives they are meant to live — to express their natural behaviors and proclivities, to be who they are and to get what they need. For example, taking wolves from one area, where they have good lives, and moving them to another where they may be killed and have lower-quality lives is an assault on their emotional lives — animals are not here for us to do whatever we want with them.

Increasing recognition of consciousness and cognition in other species has also led to efforts to grant certain species legal personhood. These efforts target nonhuman species we identify as particularly intelligent — like great apes, elephants, and whales — but I think paying attention to their emotions, feelings, and personalities also makes these even more convincing arguments. All in all, over the last two decades, it's been exciting to see so much more concern about animal emotions, human-animal relationships, and the development and implementation of nonkilling solutions when dealing with inevitable human-animal conflicts.

Personally, I have always felt that dogs are the best species to help us bridge the empathy gap and foster our sense of connection with other animals. Granted, the lives of companion animals can be problematic and stressful in themselves (many companion animals pay a high price to live with us), but they also can expand empathy to other nonhumans, as well as to humans, and help unite us all. Evidence for sentience is often experienced most vividly and directly with our dogs and other companions, and this inspires us to give them the best lives possible. This can easily spill over into more freedom and justice for all animals, including ourselves. Who could argue that more trust, empathy, compassion, freedom, and justice wouldn't be the best thing we could do for all animals and for the future generations who will inherit our wondrous planet?

Humans have enormous power to affect the world any way we choose. In our increasingly human-dominated world, we daily silence and slaughter sentience in innumerable animals — trillions in fact. However, we also know that we're not the only sentient creatures with feelings, and with this knowledge comes the enormous responsibility and obligation to treat other beings with respect, appreciation, compassion, and love. There's no doubt whatsoever that, when it comes to what we can and cannot do to other animals, it's their emotions that should inform our discussions and our actions on their behalf, and we can always do more for them. So this remains a forward-looking book of hope that stresses that we must be imaginative in our interactions with other animals.

Emotions are the gifts of our ancestors. We have them and so do other animals. We must never forget this.

⊰{ A Note about Terminology }⊱

Before I delve deeply into the emotional lives of animals, I want to address the important matter of terminology. In discussions of animal emotions, we sometimes forget that humans are also animals. However, it's cumbersome to use the phrase "nonhuman animals" to refer to beings we typically call "animals." And so whenever I use the word *animals*, I intend it to mean "nonhuman animals" — realizing of course that we're all animals. My hope is that this linguistic shorthand won't perpetuate any "forgetfulness."

In addition, when it comes to pronouns, I always use *who*, *he*, *she*, and *they* when referring to animals, and never *that*, *which*, or *it*. Animals are subjective beings, and so I use subjective pronouns. For more on language, see "Honest Talk and Rewilding Language" (pages 173–76).

{ 1 }

The Indisputable Case for Animal Emotions and Why They Matter

Anthropocentrism is at the root of all abuse of our fellow creatures on earth — the logically unsupportable belief that humans are the only species on the planet worthy of consideration.

— Sir Brian May,
founding member of Queen and the Save Me Trust

In September 2006, there was a meeting about animal welfare called "The Heart of the Matter." At the time, I was very pleased to see scientists finally using the word *heart*. Today, this is much more common. Researchers write about heart and love, and it is refreshing and essential that they do. Not only is it OK for scientists to feel for other animals, but it makes for better science. Why study animal emotions if you don't think they exist or matter to the animals themselves?

I study animal emotions and I love what I do. Over the course of my career, I've studied a wide variety of animals — coyotes, wolves, dogs, Adélie penguins, archerfish, western evening grosbeaks, and Steller's jays — and I've tackled a wide range of questions, dealing with everything from social behavior, social organization, and social development to communication, play, antipredatory behavior, aggression, parental behavior, and morality. To me, the evidence for animal emotions is indisputable and impossible to deny, and it is now widely supported by extensive research in animal behavior, cognitive ethology, evolutionary biology, and neurobiology. But most people don't need science to confirm the obvious: Many animals display their feelings openly, publicly, for anyone to see, hear, or sometimes smell. Simply put, science now confirms that what we see outside tells us lots about what's happening inside an individual's head and heart. This is important, for careful scientific research is validating what we intuitively understand: that numerous and diverse animals feel, and their emotions are as important to them as ours are to us.

In this chapter, I focus on what emotions are, what they indicate about animal minds, how those emotions impact us, and ultimately, what to do with this knowledge. In fact, that is the goal of this whole book — to show that animal emotions exist, that understanding them should be important to humans, and that this knowledge must influence how we treat other animals.

In discussing animal emotions, I focus mainly on behavioral data and anecdotal stories, weaving in recent discoveries in social neuroscience to show how a combination of common sense and scientific data — what I call "science sense" — makes a strong case for the existence of beastly passions. However, once we agree that animal emotions exist and that they matter, then what? Then we must consider ethics. We must look to our actions — at all the ways we use and abuse and treat other animals — and see if they are consistent with our knowledge and beliefs. I feel strongly that ethics should always inform science. We should always strive to merge knowledge, action, and compassion. That is always the heart of the matter.

Bird Funerals: Showing Avian Respect Isn't "For the Birds"

A number of years ago, my friend Rod and I were riding our bicycles around Boulder, Colorado, when we witnessed a very interesting encounter among five magpies. Magpies are corvids, a very intelligent family of birds. One magpie had obviously been hit by a car and was lying dead on the side of the road. The four other magpies were standing around him. One approached the corpse, gently pecked at it — just as an elephant noses the carcass of another elephant — and stepped back. Another magpie did the same thing. Next, one of the magpies flew off, brought back some grass, and laid it by the corpse. Another magpie did the same. Then all four magpies stood vigil for a few seconds and one by one flew off. Rod was astounded by how deliberate the birds were. He asked me if this was normal magpie behavior, and

I told him that I'd never seen anything like this before. At the time I hadn't read any accounts of grieving magpies. Reading their actions, though, we couldn't help but conclude that these birds were saying a magpie farewell to their friend.

Over the years since I wrote about this amazing funeral, I've received stories about magpies, sparrows, and other birds that are identical to what Rod and I saw. In their wonderful book *Gifts of the Crow*, John Marzluff and Tony Angell note, "Crows and ravens routinely gather around the dead of their own species [but] rarely do they touch the body." John sent me a story from Vincent Hagel, formerly the president of the Whidbey Audubon Society, about a crow funeral that closely resembles what we saw with the magpies. Vincent wrote:

> My good buddy and I were in his mother's kitchen as she prepared an after-school snack for us. Suddenly, she told us to quickly look out the kitchen window. Just a few feet from the house lay an obviously dead crow, and about twelve other crows were hopping in a circle around the body. After a minute or two, one crow flew off for a few seconds, then returned with a small twig or piece of dried grass. It dropped the twig on the body, then flew away. Then, one by one, the other crows each left briefly, one at a time, and returned to drop grass or a twig on the body, then fly off until all were gone, and the body lay alone with twigs lain across it. The entire incident probably lasted four or five minutes.

Here's another moving story from Hasan Jasim:

> I heard a sound at the window of the cottage. When I looked, a cedar waxwing had hit the window and fallen to the deck. Its mate stood beside it. It was clear that the fallen bird was dead. I turned away for a moment, and when I looked again, its mate had hopped over. It lay down beside its partner, their heads touching. It closed its eyes and passed over to Spirit with its mate. The sadness and beauty of the moment were such a privilege to witness. If anyone has ever doubted that animals feel emotion and love, this moment was absolute proof of both. Nature truly has much to teach us.

As observers, we are left to wonder: Were these birds thinking about what they were doing? Were they showing avian respect for their friends and mates? Or were they merely acting "as if" they cared? Were they just animal automatons? Of course, we can't know what was going on inside the heads and hearts of these particular birds, but the collective weight of similar anecdotes and the state of current research is such that I feel very comfortable answering these questions, in order: yes, yes, no, no.

Thick Skin and Tender Hearts: Babyl the Elephant

Here is another example of animal behavior that clearly expresses an animal's inner life. On a trip to Kenya and Tanzania many years ago, my eyes were opened to the world of elephants, who are some of the most amazing beings I've ever seen up close and personal. Observing large groups of wild elephants, I could *feel* their majestic presence, awareness, and emotions. These firsthand experiences rippled through my body, brain, and heart, and they were wholly different than seeing captive elephants, who often live alone or in unnatural groups while confined in the unnatural settings of a zoo. My encounters in Africa were deeply spiritual, inspirational, and transformative, and ever after, it made me deeply grateful for every effort to preserve these magnificent beings.

While we were watching a group of wild elephants living in the Samburu Reserve in northern Kenya, we noted that one of them, Babyl, walked very slowly. We learned that she was crippled and that she couldn't travel as fast as the rest of the herd. However, we saw that the elephants in Babyl's group didn't leave her behind; they waited for her. When I asked our guide, the elephant expert Iain Douglas-Hamilton, about this, he said that these elephants always waited for Babyl, and they'd been doing so for years. They would walk for a while, then stop and look around to see where Babyl was. Depending on how she was doing, they'd either wait or proceed. Iain said the matriarch even fed her on occasion.

Why did the other elephants in the herd act this way? Babyl could do nothing for them, so there seemed no reason for or practical gain in helping her. The only obvious conclusion we could draw was that the other elephants cared for Babyl, and so they adjusted their behavior to allow her to remain with the group.

Friendship and empathy go a long way. And Babyl's friends aren't an isolated example. In October 2006, in a small village in eastern India, a group of fourteen elephants crashed through a village looking for a group member who had fallen into a ditch and drowned. Residents had already buried the seventeen-year-old female elephant, but still, thousands of people were forced to flee their homes as the other elephants searched and rampaged for more than three days.

What Are Emotions?

It is difficult to answer the question "What are emotions?" Most of us know emotions when we see them but find it difficult to define them. Are they physical, mental, or both? As a scientist, I feel safe saying that emotions are psychological phenomena that help in behavioral management and control; they are phenomena that emote us, that make us move. A distinction is often made between "emotional responses" to physical reactions and "feelings" that arise from thoughts. Emotional responses show that the body is responding to certain external stimuli. For example, we see an oncoming car about to hit us and feel fear — increasing our heart rate, blood pressure, and body temperature. But actually, the fear isn't felt until the brain responds to the physiological changes that are a reaction to seeing the oncoming car.

Feelings, on the other hand, are psychological phenomena, events that happen solely in an individual's brain. An external event may trigger one emotion, such as anger or grief, but upon reflection we may decide we feel differently. We may interpret our emotions. Feelings express themselves as different moods. Feelings help us and influence how we interact with others in a wide variety of different social situations. As psychologist Rachel Allyn aptly puts it, "Emotions originate as sensations in the body. Feelings are influenced by our emotions but are generated from our mental thoughts." She notes, "Emotions are real-time data sparked by sensations in the body [whereas] feelings can be more biased, altered by mental misconceptions." Many people don't distinguish between emotions and feelings because of how closely associated they are, and it doesn't really bother me if the words are used interchangeably.

Charles Darwin, the first scientist to study animal emotions systematically, recognized six universal emotions: anger, happiness, sadness, disgust, fear, and surprise. He maintained these core emotions help us deal rapidly with a wide variety of circumstances and help us to get along in a complex social world. Others have since added to his list. Stuart Walton, in his book *A Natural History of Human Emotions*, adds jealousy, contempt, shame, and embarrassment to Darwin's core group, while the neuroscientist Antonio Damasio (in *Descartes' Error*) says that social emotions also include sympathy, guilt, pride, envy, admiration, and indignation. It's interesting that none of these researchers mention love.

Which, if any, of these emotions do animals experience? And do animals experience any emotions that humans do not? These are very interesting questions. Ethologist Joyce Poole, who has studied elephants for many years, states: "While I feel confident that elephants feel some emotions that we do not, and vice versa, I also believe that we experience many emotions in common." While I agree that it's possible that we and other animals might have unique emotional experiences, I still don't know how to cash this out without using familiar words. I always say, for example, that there's chimpanzee joy, dog joy, cat joy, and so on, and it's the same for any other positive or negative emotion. I think that words like *joy*, *grief*, and *jealousy* share a common enough root that it's valid to use them to describe what an individual of another species is experiencing. There also are individual differences within species, but that doesn't mean, say, that if your joy and grief are different from mine, I experience them and you don't, or vice versa.

If Poole is right, and there are some emotions animals experience that humans will never understand, there are many that we do. Aren't animals, human and nonhuman alike, happy when playing or when reuniting with a loved one? When wolves meet, wagging their tails loosely to and fro in a circle, whining and jumping about, are they not displaying happiness? What about elephants who reunite in a greeting celebration, flapping their ears and spinning about and emitting a vocalization known as a "greeting rumble" — is this not happiness? Likewise, don't animals become sad after losing a close friend? What name but grief can we give to the emotion that animals display when they remove themselves from their social group, sulk after the death of a friend, stop eating, and even die? Surely, despite differences, all species must share a similar core of emotions.

Primary and Secondary Emotions

Researchers usually recognize two different types of emotions, primary and secondary emotions. Primary emotions are considered to be basic inborn emotions. These include generalized, rapid, reflex-like ("automatic" or hardwired) fear and fight-or-flight responses to stimuli that represent danger. They require no conscious thought and include Darwin's six universal emotions: fear, anger, disgust, surprise, sadness, and happiness. Animals can perform a primary fear response, such as avoiding an object, almost unconsciously, before they have even recognized the object generating the reaction. Loud raucous sounds, certain odors, objects flying overhead: These and other such stimuli are often inborn signals for "danger" that cause an automatic avoidance reaction. There's little or no room for error when confronted with a dangerous stimulus, so natural selection has resulted in innate reactions that are crucial to individual survival.

Primary emotions are wired into the brain's evolutionarily old limbic system (especially the amygdala); this is the "emotional" part of the brain (so named by Paul MacLean in 1952). The physical structures in the limbic system and similar emotional circuits are shared among many different species and provide a neural substrate for primary emotions. In his three-brains-in-one (or triune brain) theory, MacLean identifies the reptilian, or primitive, brain (possessed by fish, amphibians, reptiles, birds, and mammals); the limbic, or paleomammalian, brain (possessed by all mammals); and the neocortical, or "rational" neomammalian, brain (possessed by a few mammals, such as primates and humans) — all packaged into the cranium. Each is connected to the other two, but each also has its own capacities. While the limbic system seems to be the main area of the brain in which many emotions reside, current research now indicates that not all emotions are necessarily packaged into a single system, and there may be more than one emotional system in the brain. Studies also show that humans and animals share similar chemical and neurobiological systems related to the primary emotions.

Secondary emotions are more complex emotions, and they involve higher brain centers in the cerebral cortex. They could involve core emotions of fear and anger, or they could be more nuanced, involving such things as regret, longing, or jealousy. Secondary emotions are not automatic: They are processed in the brain, and the individual thinks about them and considers what to do about them — what action is the best one to perform in a certain situation. Context is key — bared teeth or growling mean different things in different situations, as might howling or whimpering. This is important to keep in mind when observing animals: The behaviors on display can indicate different feelings depending on the context or situation, or who's involved and where they are. I can't emphasize too many times how important it is to watch animals carefully, to get to know them and consider all possibilities.

For instance, conscious thought and secondary emotions can influence how we respond to situations that bring forth primary emotions: We may duck in fear as an unseen object flies overhead, but as we recognize that it's only a shadow, we refrain from running and instead, with a twinge of embarrassment, quickly straighten up and pretend nothing is wrong. We know from our own experiences that not only are emotions complex but they can shift quickly.

Thinking about the emotion allows for flexibility of response in changing situations after evaluating which of a variety of actions would be the most appropriate to perform in the specific situation. Sometimes, if someone is bothering you, it might be appropriate to get away from them, and sometimes this might create an even worse social situation — depending on who the person is and what kind of consequences you fear. Although many emotional responses are unconsciously generated — they occur without thinking — we learn to try to think before acting. Thinking allows us to make connections between feelings and actions, and this allows for variability and flexibility in our behavior so that, depending on the social situation, we always do the right thing. In this way, evidence of emotions in any creature is also an important step in determining consciousness and self-awareness.

Dogs Are Happy, Not "Happy"

*The reason a dog has so many friends
is that he wags his tail instead of his tongue.*

— Anonymous

We've all seen it. Maddy, Zeke, and Mickey, three of my friend's dogs, regularly had playdates at my house when their human companions were away. They arrived bounding around wildly in play, panting and barking, their wagging tails seemingly propelling them through space. They tried to play with anyone who was available, whirling around to catch their own tail, running amok and knocking down anything and anyone in their way, stopping only for a taunting pause and then jumping into play once again. There's no question about it: These dogs were having fun! Sometimes they began "going crazy" in the car as they were driven up the road to my house, knowing what was waiting for them when they arrived. They knew it was going to be a fun day at Marc's place.

Most people with dogs are familiar with zoomies, when dogs zoom around with reckless abandon. Also called FRAPs, or frenetic random activity periods, zoomies aren't simply expressions of dogs suffering from ADHD — the dogs really enjoy them. Zoomies are their own reward, and for some dogs, that's all there is to it. Animals often play just for the hell of it — for fun — because it feels good.

When we take dog play seriously, and study it, we gather all the "proof" we need that animals have emotions, since dogs don't hide what they feel. The Nobel Prize–winning ethologist Konrad Lorenz gave us a very simple and common example when he noted how publicly emotional dogs are when anticipating a walk. Lorenz wrote in *Man Meets Dog*: "The owner says without special intonation and avoiding mention of the dog's name, 'I don't know whether I'll take him or not.' At once the dog is on the spot, wagging his tail and dancing with excitement.... Should his master say, 'I don't think I'll take him, after all,' the expectantly pricked ears will drop sadly.... On the final pronouncement, 'I'll leave him at home,' the dog turns dejectedly away and lies down again." I've seen this countless times among the dogs with whom I've shared my home and heart, and I am always very careful not to send any ambiguous messages — why ruin what they're rehearsing in their heads?

Thankfully, the dismissive, skeptical line that animals only act "as if" they're feeling joy, grief, anger, or pain is now essentially dead. I know no practicing researcher who doesn't attribute emotions to their companion animals — who doesn't freely anthropomorphize — at home or at cocktail parties, regardless of what they do at work. This anthropomorphizing is nothing to be ashamed of, by the way; as Alexandra Horowitz and I have argued, and as I explain in chapter 5, these scientists are simply doing what comes naturally. Anthropomorphizing is an evolved perceptual strategy; we've been shaped by natural selection to view animals in this way. The only thing that's really changed is that, now that many scientists largely accept as fact that animals share a wide range of emotions, they have caught up to what most people have recognized all along.

One example of how scientists accept the universality of the primary emotions is the use of animals to develop and test drugs to treat human mental disorders. Mice are considered a good model for sadness and introversion. After mice are bullied or consistently dominated by other mice, they become withdrawn, and these depressed mice respond to human drugs such as the antidepressant Prozac. In another example, suicidal rats — or rats who have toxoplasmosis and develop a suicidal attraction to cats — can be successfully treated with antipsychotic drugs. When given haloperidol, which is used to control schizophrenia, their fondness for cats decreases greatly. The veterinarian Nicholas Dodman suggests using similar drugs along with behavioral conditioning for problem dogs and cats. If animals respond to these drugs as humans do, then it's highly likely that they have similar neural underpinnings to their emotions and probably similar feelings.

Empathy or compassion is an important secondary emotion to identify in animals, for it demonstrates a selfless caring for others. Recall Babyl and her caring friends. While I was in Homer, Alaska, I read a similar story about two grizzly bear cubs who stuck together after they were orphaned when their mother was shot near the Russian River. The female cub remained with her wounded sibling, though he limped, swam very slowly, and needed help to get food. An observer noted, "She came out and got a fish, and pulled it back, and then she let the other one eat." The young female obviously cared for her brother, and her support was crucial for his survival.

There's also a story of a troop of about a hundred rhesus monkeys in Tezpur, India, who brought traffic to a halt after a baby monkey was hit by a car. The monkeys encircled the injured infant, whose hind legs were crushed and who lay in the road unable to move, and blocked all traffic. A government official reported that the monkeys were angry, and a local shopkeeper said: "It was very emotional.... Some of them massaged its legs. Finally, they left the scene carrying the injured baby with them."

In one classic study, a hungry rhesus monkey would not take food if doing so subjected another monkey to an electric shock, and there is a more recent scientific study on empathy in mice. In this study, either one or both members of a pair of adult mice were injected with acetic acid, causing them to writhe in pain, so that researchers could observe whether or not these rodents had the capacity to feel for others in pain. Researchers discovered that mice who watch their peers in pain are more sensitive to it themselves and that an injected mouse writhed more if its partner was also writhing. The mice used visual cues to generate the empathic response, although they typically use scent in many of their social encounters. Two other studies showed that rats would free trapped rats and even forgo eating chocolate if they saw another rat drowning. Not only do these stories show that mice, and many nonhuman animals, possess empathy, but we know that the empathic response in mice is mediated by the same brain mechanisms as in human empathy.

Of course, these studies are troubling. Did the scientists really need to cause such wrongful pain to reach their conclusions? What does it say about us that we chose to be cruel in order to prove that other animals can be empathetic and compassionate? In fact, these studies on mice and rats might not have been allowed if they had been protected by the Animal Welfare Act. When I wrote this book seventeen years ago, I'd hoped by now that solid science would have led to changes in the laws governing research. As I discuss in chapter 6, we're still waiting.

I continue to receive numerous stories about empathy in a wide variety of animals, including rodents. People who live with animals aren't surprised by the scientific findings. CeAnn Lambert, who worked at the Indiana Coyote

Rescue Center, told me that one hot summer morning she saw two baby mice in a deep sink in her garage. They were trying to get out of the sink but couldn't get up the steep, slick sides. One seemed less exhausted than the other. CeAnn put some water in a lid and placed it in the sink, and immediately the more lively pup went over to get a drink. On the way to the water the mouse found a piece of food and took it over to their littermate. The weak mouse tried to take a bite while the other kept moving the food slowly toward the water. Finally, the weaker mouse got a drink. Both gained some strength and climbed out using a board that CeAnn put in the sink.

A wonderful example of cross-species empathy and compassion was reported in September 2023: After a stray dog jumped into the River Savitri in India while being chased by a group of feral dogs, three huge mugger crocodiles pushed the dog to safety using their snouts, when they surely could have enjoyed an easy meal.

Ultimately, the point is, even if animal emotions aren't exactly the same as our own, or for that matter the same across species — and there is no reason to think they are or have to be — this doesn't mean that animals don't feel. In fact, the rich emotional lives of animals seem to indicate that they are also engaged in a good deal of conscious thought.

Animal Feelings Are Transparent

> *The reluctance of contemporary philosophers and scientists to embrace the view that animals have minds is primarily a fact about their philosophy and science rather than a fact about animals.*
>
> — DALE JAMIESON, "SCIENCE, KNOWLEDGE, AND ANIMAL MINDS"

When animals bark, howl, purr, whimper, grunt, laugh, or squeal, it means something to them, and what they're saying should also mean something to us, for their feelings matter. Lynne Sharpe points out in her wonderful book *Creatures Like Us?* that the interests and concerns of animals are as important to them as ours are to us. Tails talk to us about what animals are feeling, and so, too, do various postures, gaits, facial expressions, sounds, and odors. Sometimes I wish I had a tail and mobile ears so I could communicate more effectively with dogs and other animals, whose tails and ears tell us lots about what they're thinking and feeling. Wagging wildly or drooping between their legs, animals' tails allow us to enter into their own brand of sentience.

What animals know — and how much self-awareness they have — is a topic of wide and often heated debate. The growing scientific evidence is that they know quite a lot, but the difficulties of communicating across species may make it nearly impossible to ever know exactly how much. My baseline concerning animal emotions and sentience is pretty simple — animals will always have their secrets, but their emotional experiences are transparent. In other words, we know that numerous animals feel a rich panoply of emotions, some of which, like empathy, require a certain level of conscious thought. Many animals display a sense of humor. A few animals, such as chimpanzees, dolphins, elephants, magpies, and fishes (including manta rays and bluestreak cleaner wrasses) have passed mirror tests that demonstrate they possess some form of self-awareness. Some animals might experience a sense of awe, and some might even be moral beings who know "right" from "wrong."

Of course, there are differences among species. We would expect variations based on social, ecological, and physical factors. However, there are compelling similarities despite sometimes extreme differences. One common measuring stick is called "relative brain size" (brain size expressed as a ratio to body size), and indeed, just about all researchers agree that, when comparing species, relative brain size makes a difference in various aspects of behavior, including antipredatory and feeding strategies. Just what these types of differences mean, however, remains largely a mystery, but there's no evidence that it means that animals with a smaller ratio don't have rich emotional lives. Because we share old parts of the brain that are important in human emotions, namely the limbic system, including the almond-shaped structure called the amygdala, focusing solely on relative brain size is misleading. We need to pay attention to what we share with other animals and not necessarily how much of it we share with them. The brains of mice, dogs, elephants, birds, and humans differ greatly in size, but all of these species display joy and empathy.

Australian biologist Gisela Kaplan learned that birds who engage in social play tend to have larger brains than birds who don't. There's no reason not to assume they enjoy playing, that it feels good — why else would they do it? Even nonsocial playing birds had larger brains than nonplayers. Large-brained birds also tend to live longer than birds with smaller brains.

Unfortunately, misconceptions continue, often in popular books that offer unsupported generalizations about the cognitive, emotional, and empathic capacities of animals. For example, Harvard psychologist Daniel Gilbert, in his popular bestseller *Stumbling on Happiness*, claims that "*The human animal is the only animal that thinks about the future*" (Gilbert's italics) and that this "is a defining feature of our humanity." Not even the animal emotion skeptics

I know would ever dare to make this erroneous claim; there are literally volumes of data showing that individuals of many species think about the future, from Mexican jays, red foxes, and wolves caching food for later retrieval to a subordinate chimpanzee or wolf pretending that they don't see a favored food item in the presence of a dominant individual and later returning to eat it when the dominant animal isn't around.

Another example is Gerald Hüther's claim, in his book *The Compassionate Brain*, that the capacity for empathy sets the human brain apart from all other nervous systems, despite scientific evidence that this just isn't so. Such anthropocentric arrogance — human exceptionalism — is both inexcusable and damaging to other animals. Some people still don't know what's happening in the study of nonhuman minds or write off animal sentience and intelligence as science fiction. But it's not that difficult anymore to dethrone humans from feeling separate from and above all other animals.

In the end, the truth is simply that dogs have rich emotional and cognitive experiences of the dog kind, and the same can be said for many other animals. Ethological studies and research in social neuroscience show that humans aren't in sole possession of any particular emotional or mental ability. No, animals can't speak to us. They let their faces, eyes, ears, and tails do the talking; some communicate with odors, ranging from sweet to pungent. Animals also convey what they feel and think using a myriad of behavior patterns — postures, gestures, and gaits. In all these ways, animals express themselves, but are we listening?

Welcome to the Sentience Club: Insects, Crustaceans, and Plants

In my musings about animal sentience and emotions, I can't help wondering, What about insects? What are their lived experiences? Do mosquitoes have emotional lives? What about ants, bees, and beetles? What about fishes, shrimp, and crabs? What about *plants*?

The more research that's done, the more surprises we're finding, and the more beings we're welcoming into the ever-growing sentience club. Of course, one of the first things people wonder is, aren't insect brains too tiny? They lack the same neural apparatus as mammals and other animals, and they lack facial expressions and most of the other physical behaviors through which we readily recognize emotions, so it's difficult to know. But researchers are figuring out ways to recognize sentience in species we'd never expect.

It's known that bees can get depressed, feel pain, and be optimists or pessimists. Bees have small but stunning brains and are surely sentient. In a detailed 2022 review called "Can Insects Feel Pain? A Review of the Neural and Behavioural Evidence," Matilda Gibbons and her colleagues discovered that flies, mosquitoes, cockroaches, and termites showed strong evidence of pain, and beetles, butterflies, and moths showed substantial evidence of pain. Importantly, no insects failed any of the criteria that were used to assess their capacity for feeling pain.

Not only is this amazing and fascinating, but these data are essential to use in management, farming, and research and to help us understand the true extent of the biodiversity of sentience. Biologist Meghan Barrett and her colleagues are deeply concerned about the mistreatment of insect pollinators, and they argue that strict animal welfare standards should be applied to conservation efforts that focus on them. For example, lethal sampling methods are still used to sample their populations, and they argue that the welfare of *individual* pollinators may be morally significant. I agree, and the focus on individuals is consistent with the goal of compassionate conservation, in which the life of every individual matters and is morally significant.

In their study on insects, Gibbons and her colleagues relied on a seminal essay by Jonathan Birch and his colleagues called "Review of the Evidence of Sentience in Cephalopod Molluscs and Decapod Crustaceans." This essay laid out eight criteria to assess sentience in cephalopod mollusks (squid and octopus) and decapod crustaceans (shrimp, lobsters, crayfish, and crabs), and then they conducted research for evidence. Here are the eight criteria (which I have modestly edited):

1. The animal possesses receptors sensitive to noxious stimuli (nociceptors).
2. The animal possesses integrative brain regions capable of integrating information from different sensory sources.
3. The animal possesses neural pathways connecting the nociceptors to the integrative brain regions.
4. The animal's behavioral response to a noxious stimulus is modulated by chemical compounds affecting the nervous system in either or both of the following ways: (a) The animal possesses an endogenous neurotransmitter system that modulates its responses to threatened or actual noxious stimuli; or (b) putative local anesthetics, analgesics, anxiolytics, or antidepressants modify an animal's responses to threatened or actual noxious stimuli.

5. The animal shows motivational trade-offs, in which the disvalue of a noxious or threatening stimulus is weighed against the value of an opportunity for reward, leading to flexible decision-making. Enough flexibility must be shown to indicate centralized, integrative processing of information involving an evaluative common currency.
6. The animal shows flexible self-protective behavior (wound-tending, guarding, grooming, rubbing) of a type likely to involve representing the bodily location of an injury or noxious stimulus.
7. The animal shows associative learning in which noxious stimuli become associated with neutral stimuli, and/or in which novel ways of avoiding noxious stimuli are learned through reinforcement.
8. The animal shows that it values a putative analgesic or anesthetic when injured in one or more of the following ways: (a) the animal learns to self-administer putative analgesics or anesthetics when injured; or (b) the animal learns to prefer, when injured, a location at which analgesics or anesthetics can be accessed; or (c) the animal prioritizes obtaining these compounds over other needs (such as food) when injured.

Research focused on specific types of octopuses, squid, shrimp, crabs, lobsters, and nautiluses, and the study concluded, "We recommend that all cephalopod molluscs and decapod crustaceans be regarded as sentient animals for the purposes of UK animal welfare law. They should be counted as 'animals' for the purposes of the Animal Welfare Act 2006 and included in the scope of any future legislation relating to animal sentience."

For some time, scientists have recognized that fishes experience pain and fear. Based on his own research, world-renowned scientist Ian Duncan pronounced this at the 2006 World Society for the Protection of Animals conference in Rio de Janeiro. Fishes are also cunning, deceitful, and display cultural traditions. In her classic 2010 book *Do Fish Feel Pain?*, fish expert Dr. Victoria Braithwaite concluded, "I have argued that there is as much evidence that fish feel pain and suffer as there is for birds and mammals — and more than there is for human neonates and preterm babies."

Knowing that animals are sentient should — and must — make a difference in how we view, represent, and treat them. Who'd have thought that small-brained flies, mosquitoes and moths, squid and crabs, would show strong evidence of pain? In fact, they do, and the details — the breadth and depth — of this research set an excellent example for future comparative research on many other animals. I look forward to seeing what these studies yield about the biodiversity of sentience, and I'm sure that more "surprises" will be uncovered, such as...

What about plants and other flora? Recently, there has been growing interest in plants, with serious, card-carrying scientists studying plant intelligence, emotions, and sentience. Some suggest that plants do feel pain and may even "cry" when they're stressed or injured. Likewise, just as ample data show that many people find that animals bring them comfort and make them feel better, so, too, do plants. What's interesting to me, coming from the "animal" side of things, is that much of what is currently being discussed mirrors the debates that occurred decades ago when people wondered if *animals* were conscious, emotional, and sentient, and they wanted to know how we could "prove" it. And following a similar historical course of inquiry that characterized the growing interest in animal minds, many scientists studying plants have moved on from asking *if* plants are smart and emotional to asking instead *why* these capacities have evolved.

As someone who has studied nonhuman animal minds for decades, I've seen the narrow and dismissive views about animal sentience change radically. Where once people questioned whether any nonhumans really had minds, now people accept that many if not most animals do. I'm sure that we'll see a similar broadening of attitudes about plant minds as relevant studies provide the "proof" and as people realize that sentience is not rare or exclusive but a defining aspect of life itself. Closing the door on learning about the "biodiversity of sentience" only robs us of learning about its taxonomic range.

Animals and Humans: Emotions Are Social Glue Underlying Empathy

Animal emotions are important for many reasons, but one is because we need animals in our lives; they help us. It's because animals have emotions that many people are so naturally drawn to them; lacking a shared language, emotions are perhaps our most effective means of cross-species communication. We can share our emotions, we can understand the language of feelings, and that's why we form deep and enduring social bonds with many other beings. Emotions are the glue that underlie empathy. They catalyze and regulate social interactions in animals and in humans. Pet rocks and stuffed or robotic animal toys don't have the same effect.

Shared emotions and their glue-like adhesive power are responsible for this country's billion-dollar pet industry. In 2022, around 70 percent of US households had a least one companion animal, about 70 million had a dog, and about 45 million had a cat. But the variety of animals kept as pets, particularly worldwide, is astonishing; it includes rodents, birds, fish, amphibians, reptiles, insects, spiders, invertebrates, and many more. About 20 percent of US households have a bird, and more than 600 million pet fish are sold each year. In both the United States and Britain, the numbers of pets are growing.

Veterinarian Marty Becker's book *The Healing Power of Pets* shows how pets can keep many people healthy and happy — they help to heal lonely people in nursing homes, hospitals, and schools. In *Kindred Spirits*, holistic veterinarian Allen Schoen points to fourteen concrete ways in which a relationship between animal companions and humans has been shown to reduce stress. These include reducing blood pressure, increasing self-esteem in children and adolescents, increasing the survival rate of heart attack victims, improving the life of senior citizens, aiding in the development of humane attitudes in children, providing a sense of emotional stability for foster children, reducing the demand for physician's services for nonserious problems among Medicare enrollees, and reducing loneliness in preadolescents. And Michelle Rivera, in her book *Hospice Hounds*, tells numerous stories of how dogs and cats can help people who are near death.

One study showed that a visit from a friendly pup might be good medicine for an ailing heart. In a randomized study of seventy-six hospitalized heart-failure patients, UCLA researchers found that anxiety scores dropped an average of 24 percent among patients who interacted with canines, regardless of the breed. The dogs would lie on the patients' beds for twelve minutes while patients simply patted and scratched their ears. "This study demonstrates that even a short-term exposure to dogs has beneficial physiological and psychosocial effects on patients who want it," said Kathie Cole, a clinical nurse at the UCLA Medical Center.

Similarly, in my home state, inmates at the Denver Women's Correctional Facility get to care for and live with dogs who would have been put to sleep at the local animal shelter. The experience of walking the dogs, grooming them, and cleaning up after them is incredibly rewarding and beneficial to everyone — the inmates, the dogs, and the prison staff. One woman I knew went on to found a dog training business after her release.

That said, animals are not necessarily a global panacea for every human ailment. I am fully aware that nonhumans don't *always* help humans. We know — from scientific studies, personal stories, and common sense — that for some people, canine and feline companions, or any animal in the home, can cause more stress and anxiety and not necessarily make life "easier."

Psychologist Hal Herzog analyzed twenty-one studies and found that the majority did not support the conclusion that people living with pets are less lonely than people without pets; in fact, only one of the eight studies conducted since 2015 found strong evidence that pets reduce loneliness. Pets also don't always do a very good job of relieving depression. Nonetheless, when people are making personal decisions about their own lives, my advice is always to do what seems best for both humans and nonhumans, regardless of what studies or other people say. Living with anyone is a two-way street, and if it's not good for *everyone* — if it's not a win-win for all — it's best not to take on the responsibility. Animal companions provide comfort and happiness for some, and they don't for others, so let your life and heart, and the needs and wants of the animal, be your guide.

Stories of wild animal and human encounters — and other cross-species relationships — also provide evidence of the power of emotional bonds. Lions are magnificent carnivores and powerful predators, yet they also show compassion, sympathy, and empathy in very unpredictable ways. For example, three lions in Ethiopia rescued a twelve-year-old girl from a gang who had kidnapped her. Said Sergeant Wondimu Wedajo: "They stood guard until we found her and then they just left her like a gift and went back into the forest." Stuart Williams, a wildlife expert with the country's rural development ministry, said that it was likely that the young girl was saved because she was crying from the trauma of her attack. Williams said, "A young girl whimpering could be mistaken for the mewing sound from a lion cub, which in turn could explain why [the lions] didn't eat her. Otherwise they probably would have done so." Eventually, three of the four kidnappers were caught.

In but one of many stories of dolphins helping humans at sea, in New Zealand, a pod of dolphins circled protectively around a group of swimmers to fend off an attack by a great white shark. "They started to herd us up. They pushed all four of us together by doing tight circles around us," said Rob Howes, one of the swimmers. In these stories, we see that the empathetic presence of animals can have a direct and immediate impact on our well-being and even survival.

If it seems strange that animals would go out of their way to care for us, that's not even half the story. Some of the relationships that animals form are so improbable they defy belief. For instance, in the Samburu Reserve in northern Kenya, a lioness adopted a baby oryx, usually a lion's favored meal, on five different occasions. And at Tokyo's Mutsugoro Okoku Zoo, a rat snake named Aochan befriended a dwarf hamster named Gohan. The hamster was originally offered as a meal; Aochan had refused to eat frozen mice, and zookeepers figured that the live hamster would be more appetizing. However, Aochan refused to eat the animal and seemed to prefer sharing a cage with her; Gohan even napped on Aochan's back. Even though Aochan began eating frozen rodents again, he still showed no interest in eating his friend. Kazuya Yamamoto, a keeper at the zoo, said, "Aochan seems to enjoy Gohan's company very much." Wonderful "odd couple" stories pop up all the time. After they were rejected by their parents on a UK farm, Larry the lamb and Dee, a duckling, became best friends and comforted and groomed one another. When they were apart, Larry bleated and Dee quacked until they found each other again. Similarly, Joan Baez told me the lovely story of a disabled one-legged chicken she called Ms. Lopsided and her best friend, Evil Bunny, both residents of her home. The two hung out together, and when Ms. Lopsided died, Evil Bunny was very sad and moped around, clearly missing her. To help him along, Joan rescued five chickens, and Evil Bunny recovered from the loss of his close friend. If these stories don't rock your heart, little else will.

When animals express their feelings, they pour out like water from a spout. Animals' emotions are raw, unfiltered, and uncontrolled. Their joy is the purest and most contagious of joys and their grief the deepest and most devastating. Their passions bring us to our knees in delight and sorrow. If animals didn't show their feelings, it's unlikely that people would bond with them. We can form close relationships with any animal — domestic or wild, strangers or longtime companions — not only because of our own emotional needs but also because of our recognition of theirs.

Special Relationships: Best Friends and Family

I have worked with youngsters, seniors, and inmates in Jane Goodall's Roots & Shoots program for many years. Unfortunately, the Covid-19 pandemic brought many programs to a halt, and they are only slowly being rekindled as it's safe to do so. The purpose of Roots & Shoots is to stimulate children to develop respect for animals, people, and the environment. This isn't difficult: Children are curious naturalists who easily bond with all sorts of creatures.

Children also provide some of the best examples of the powerful effect of animal emotions and empathy on human lives. It's estimated that around 40 percent of children begin life in a family with domestic animals, and as many as 90 percent of all children live with a companion animal sometime during their childhood. In fact, children are more likely to grow up with a pet than with both parents, and American boys are more likely to care for pets than for older relatives or younger siblings. A vast majority of children refer to their pets as "family" or "special friends" and confidants, and more than 80 percent refer to themselves as their pet's mother or father. If stranded on a desert island, more than half of children would prefer the company of their pet rather than of family members. Children also worry about homeless pets.

Research and studies show that living with dogs or cats can improve a child's self-esteem and self-confidence and promote prosocial behaviors. When children develop positive, emotional bonds with companion animals, this helps foster empathy for others, in addition to fostering all the typical life skills children learn by helping take care of another being.

Pets also can be social catalysts and help to draw out autistic and socially withdrawn children. The term "pet therapy" was coined by Boris Levinson six decades ago, and it is still in use today. An American child psychologist, Levinson found that many children who were withdrawn and uncommunicative would come out and interact positively if his dog, Jingles, joined them in therapy sessions. Pets can help victims of psychological abuse by teaching them about unconditional love and buffering and overcoming trauma, and pets can provide support for children who have been sexually abused or have to overcome divorce or the illness or loss of a family member or close friend.

I am never at a loss for stories about kids and animals, and they always touch my heart. At one event in Denver, I met a young girl, Kory, who not only got her school to begin recycling, but who loves animals and told me how she was taken in by Agatha, a young goat being cared for by Broken Shovels Farm Sanctuary.

Perhaps the most heartbreaking story about a young girl who dreamed of becoming an animal sanctuary caretaker is that of Catherine Violet Hubbard. Catherine would "whisper to animals to tell their friends that she was nice, in hopes that they'd come in droves to visit her." But tragically, Catherine was one of the victims of the Sandy Hook shooting on December 14, 2012. To honor her daughter's dream, her mother, Jenny Hubbard, broke ground on the Catherine Violet Hubbard Animal Sanctuary exactly ten years after Catherine's death. This nonprofit sanctuary will "foster the bond between humans and animals. The sanctuary will provide veterinary care, educate visitors, and serve as a migration space for hummingbirds, bees, and other pollinators." In all honesty, I had to take a deep breath and step away from my computer and remain silent for a while as I processed this very special tribute.

As we know, it's not just children who make intimate connections with animals and draw profound strength from the animal-human bond. In my classes at the Boulder County Jail, I've had numerous discussions with inmates about their relationships with all sorts of animals, ranging from household companions to wild species they've encountered in nature. In 2015, Jane Goodall spent part of a day with my students and me talking about the wide variety of nonhumans who have helped all of us along the way. Frankly, for some of the guys, animals were their only trusted friends. In my classes, they express these feeling by writing essays and poems, some of which have been published, and their drawings have won awards. One inmate and superb artist described how learning about moon bears and how they have been able to heal from a life spent in cages had changed his life and given him hope. He wrote, "This is my spirit animal — the moon bear. I came to know that regardless of how some of these bears are treated in captivity, even they have a place in the world. I, too, have spent much of my life in captivity and confinement. I've done a lot of damage in this world, but deep within, I've known all along I have a place, and it is my calling to be part of something bigger than just me."

The value of animals to humans cannot be overstated. It's their emotions that draw us to them, and it's our emotions that likely draw them to us. And yet, while we need animals, many animals would surely do much better without us. This was true when I wrote the first edition of this book, and it rings even more true now.

A Paradigm Shift: Rethinking Our Assumptions, Revising Our Stereotypes

Questions about animal emotions and why they matter have always generated a lot of heat. Our relationship with animals is complex, and how we treat animals often changes dramatically depending on the context. Many people can show tremendous love and devotion to their household companion animals, but then, with little forethought, concern, regret, or compassion, they may abuse animals in different settings in egregious ways. For instance, when people claim to love animals and then go out and hunt them for fun and "recreation." It's also true of scientists and the different ways they treat animals at home and in the lab. My response to scientists (and others) who say they love animals and then, directly or indirectly, subject them to intentional pain and suffering is to say that I'm glad they don't love me! Unfortunately for animals, their relationship with humans has been, and remains, strongly asymmetrical. Human interests almost always trump animal interests.

Many years ago while reading the prestigious journal *Science*, I came across the following sentence: "More than any other species, we are the beneficiaries and victims of a wealth of emotional experience." The scientist who wrote this, Professor R. J. Dolan, could not possibly know that this was true. Indeed, other animals might actually experience more vivid emotions than we do, both positive and negative. This sort of "humanocentrism" still plagues the study of animal emotions, and it's also a large reason why animals are treated by such varying standards. Why are we so special? Why are we such deeply feeling animals, whereas other animals aren't? Why do we — or at least some of us — feel comfortable talking about degrees of sentience or feeling and self-servingly place humans on the top rung of the "feeling ladder"? Looking at the state of the world today, I find it difficult to accept that we should be the standard against which other animals should be compared.

We need to move on from arrogant self-centered views that mistakenly place us on top of and separate from nonhumans. Years ago, the first edition of Peter Singer's *Animal Liberation* helped me make a career-changing decision to establish a more egalitarian biology — to work hard against speciesist studies and explanations of the cognitive and emotional capacities of other animals. Since then, what we've learned about animal minds has made it painfully clear that animals are not disposable, simpleminded, unfeeling objects, and the humancentric arrogance behind the dismissive attitude "we can do anything we want to animals and nature" needs to be shelved, as it should have been decades ago. When we ignore and disrespect animals and their emotional lives, not only do our uncompassionate actions and attitudes harm them and cheapen their lives, but they damage our own integrity.

It's still my hope that the study of human-animal interactions will put an end to useless dualisms such as "us" versus "them," the "laboratory" (where animals are often disposable objects) versus "home" (where animals are highly valued friends), and "higher" animals versus "lower." These dualisms aren't accurate, and they surely do not foster the development and maintenance of deep, respectful, and symmetrical interrelationships between humans and other animals. There is no evidence that there are degrees of sentience, that so-called "smarter" animals suffer more than "dumber" animals, as philosopher Jeremy Bentham aptly noted hundreds of years ago. An individual's joy and pain are *their* joy and pain. Trying to figure out how happy or sad they might be is an interesting intellectual exercise, but we really can't know with any degree of accuracy or make valid comparisons. I know some people will disagree, but I don't find these questions to be very interesting or meaningful. Further, it's this same egregious argument that there are "higher" and "lower" species and degrees of sentience that justifies mistreatment and abuse of certain species — like rats, mice, birds, fishes, invertebrates — and respect for others, particularly charismatic mammals like dogs, dolphins, and so on.

What I've been working on for what seems to be forever is fostering a paradigm shift in how we think about animals, how we study animal emotions and animal sentience, and what we do with the information we have, "scientific" and otherwise. This paradigm shift involves revising our timeworn stereotypes about what the emotional lives of animals are "supposed" to be like. Rather than presuming that fishes feel less than mice and that mice feel less than chimpanzees, or that rats aren't as emotional as dogs or wolves, or in general that animals feel less (and know less and suffer less) than humans, let's assume that numerous animals do experience rich and deep emotions and do suffer all sorts of pain, perhaps even to a greater degree than humans.

As I explore above, we now recognize that fish, invertebrates, and even insects experience pain and fear. Furthermore, Donald Broom, a professor at Cambridge University in England, suggested the possibility that animals with more complex brains might deal more effectively with pain than animals with less complex brains because the former have more varied responses and more flexible behavior to cope with aversive situations. Broom's intriguing hypothesis is that perhaps fishes cannot deal with pain as effectively as animals with more complex brains, and because of this, fish actually suffer *more*. When deciding what and how much animals feel, it's best to keep an open mind.

As I've said, when it comes to the sometimes unconscious double standard that people frequently have in the treatment of animals, I find the question "Would you do it to your dog?" to be a great leveler. It brings home our speciesist inconsistencies — if you wouldn't do something to your dog, why would you do it to any other being?

This paradigm shift would also change how we do science — it would lead to revisions in methods and changes of heart. The burden of proof would permanently shift to the side of the skeptics, who would have to "prove" their claims that animals don't experience emotions and don't really feel pain. This is now only possible by ignoring solid science and at the same time closing one's eyes, ears, and hearts to what other animals clearly express — their thoughts and feelings. It is no longer acceptable to say, "Since we really don't know what animals feel, let's assume whatever they feel, if anything, doesn't matter." If scientists changed how they conducted experiments and tests, it would create a more humane environment for everyone. Respecting, protecting, and loving animals doesn't compromise science, nor does it mean respecting, protecting, and loving humans less. Feeding our dog doesn't require starving our children. With a little consideration and forethought, everyone can be cared for.

Most important, assuming that animals do experience rich emotions will never cause any harm. A lovely, unidentified quotation captures this well: "If I assume that animals have subjective feelings of pain, fear, hunger, and the like, and if I am mistaken in doing so, no harm will have been done. But if I assume the contrary, when in fact animals do have such feelings, then I open the way to unlimited cruelties.... Animals must have the benefit of the doubt, if indeed there be any doubt."

What We Do with What We Know

Knowing that animals feel — and being able to understand them when they express joy, grief, jealousy, and anger — allows us to connect with them and also to consider their points of view when we interact with them. Knowledge about animal passions should make a difference in how we view, represent, and treat our fellow beings.

As a scientist, I have often been criticized for being antiscience because of my strongly pro-animal views. I'm *not* antiscience. It is in the best traditions of science to ask questions about ethics; it is not antiscience to question what we do when we interact with other animals. Ethics enriches our views of other animals, as they are in their own worlds and as we relate to them in ours; they help us to see that their lives are worthy of respect, admiration, and appreciation. Indeed, it is out of respect, admiration, and appreciation that many humans seek out the company of whales, dolphins, polar bears, birds, invertebrates, and even insects.

In essence, the paradigm shift I'm arguing for is to respect each individual animal based on their inherent or intrinsic value — that is, to treat every being's life as valuable simply because they are alive — and not to distinguish and treat animals differently based on their instrumental value — or what benefits they can provide for us. All animals, humans included, want to live lives that are pain-free and filled with as much joy and happiness as possible, and it's only fair that we strive to allow this. These guiding principles, along with a focus on *individual* lives, underlie the foundation of compassionate conservation (which I discuss in chapter 6). Compassionate conservation came of age around 2010, although its roots were laid some years ago. Jessica Pierce and I formalized the science of animal well-being in our book *The Animals' Agenda*, in which we also stress that we need to bridge the "knowledge translation gap." That is, we have to use what we know on behalf of other animals, even if this is difficult and requires significant changes, rather than continuing to make facile excuses that, well, something should be done by someone at some point but "we'll get around to it later." As I mention, revising the US Federal Animal Welfare Act so that it defines the word *animal* to include rats, mice, birds, fishes, and other animals used in research is at the top of my agenda. That this still hasn't been changed is idiocy. Maybe someday soon it will be, with the help of the growing number of scientists — including Lawrence Hansen, John Gluck, and Garet Lahvis — who cannot abide this travesty.

In my own field, I know that solid science can easily be done with ethics and compassion. There's nothing wrong with compassionate or sentimental science or scientists. Studies of animal thought, emotions, and self-awareness, as well as behavioral ecology and conservation biology, can all be compassionate as well as scientifically rigorous. Science and the ethical treatment of animals aren't incompatible. We can do solid science with an open mind and a big heart.

I encourage everyone to go where their hearts take them, with love, not fear. If we all travel this road, the world will be a better place for all beings. Kinder and more humane choices will be made when we let our hearts lead the way. Compassion begets compassion and caring for and loving animals spills over into compassion and caring for humans. The umbrella of compassion is very important to share freely and widely.

Jasper the Moon Bear: Individuals Matter

> *Every man and living creature has the sacred right to the gladness of spring.*
>
> — Leo Tolstoy

I dedicated the first edition of this book to two awesome nonhumans, Jasper and Pablo. Pablo was a captive and mistreated chimpanzee, who was known as CH-377, in the New York University lab where he was kept. Jasper was a moon bear who was formerly kept in a crush cage on a bear bile farm in China. While I could have dedicated this book to the billions of animals who are abused each year, I chose these two because individuals matter. A billion animals is a statistic, but learning about and loving one animal can change minds and hearts. Here is Jasper's story:

Crush cages are still used in China. They compress a bear's body to maximize the amount of bile the animal produces (which is extracted by means of a catheter inserted in the gall bladder). Jasper was kept in a tiny cage — a "rusting prison of torture," according to Dr. Jill Robinson, the founder and CEO of Animals Asia — and tortured repeatedly over the course of many years for his bile, a chemical used in traditional Chinese medicine. Robinson wrote to me:

> This poor bear had been pressed down by a "crush" which had reduced the height of the cage by half and had flattened Jasper to the floor. Unable to sit or stand or hardly move, it is beyond belief to comprehend that this wild, intelligent bear lay there in this state for fifteen years before being rescued. Jasper was the victim of catheter implantation and his physical and mental agony must have been intolerable. A mischievous, fun-loving bear today, Jasper became everybody's friend, bears and people alike. In fact, he became a "photo-bomber" and always showed up when pictures were taken of the bears in the rescue center. His beautiful, trusting eyes showed the absolute forgiveness of his species, and reinforces our goal of rescuing as many individual bears as possible.

Most fortunately, because of the hard work of Jill and countless others, the situation for bile bears has improved over the past eighteen years, but there still is a lot of work to end bear bile farming once and for all. The organization Animals Asia is leading the way to free all bile bears from being unrelentingly tortured. I made many visits to the Moon Bear Rescue Centre outside of Chengdu, and I had the pleasure and honor of ringing the 9 a.m. feeding bell and watching the bears emerge from their cages into grassy fields to feed. For some, this transformed into a free-for-all of playing that brought tears to my eyes.

What animals feel is more important than what they know. IQs don't matter. It's worth recalling the utilitarian philosopher Jeremy Bentham's well-known statement concerning animal suffering: "The question is not, Can they reason? nor, Can they talk? but, Can they suffer?" For Bentham, it really didn't much matter if animals could think or if they were smart. Rather, Bentham was concerned with whether or not animals could suffer. Intelligence and suffering are not necessarily correlated, and clever animals don't suffer more than less-clever individuals. Some skeptics argue that some animals might not have a well-developed sense of self. We'll see this isn't really the case, but even if animals don't know who they are, they can still suffer, they can still be aware of their feelings, and they can still clearly tell us and other animals what they want and what they don't want.

When we deny that animals have feelings, it demeans both them and us. We can make their lives better with little effort other than accepting them for *who* they are — rather than viewing them only for what they can do for us — and welcoming them into our world. We should do no less.

Marc ringing the morning bell at the Moon Bear Rescue Centre, telling the bears it's time for breakfast. *(Photo credit: Rainbow.)*

Jasper in the crush cage in which he was imprisoned for fifteen years. *(Photo courtesy of Annie Mather / Animals Asia.)*

A rehabilitated Jasper.
(Photo courtesy of Annie Mather / Animals Asia.)

{ 2 }

Cognitive Ethology: Studying Animal Minds and Hearts

What we believe Equus asinus *most prefers is simply to be left alone so that they may graze casually, marvel at their surroundings, meditate on other life forms, drink plenty of water, have fun, sing, sleep, make love, raise their young, have parties, and discuss the great issues of life.*
— MICHAEL TOBIAS AND JANE MORRISON, *DONKEY*

What's a purring cat thinking or feeling? What's running through a dog's mind as they're running and playing? What's happening in an elephant's brain and heart as they nuzzle a dead mate — or a magpie's or a prairie dog's? What's a donkey feeling as they quietly graze and take in their surroundings?

I decided to become a cognitive ethologist because I want to know what swims around in the minds and hearts of other animals. To do this, I've had to learn how to see other animals for who they actually are, how to listen to "face, eye, ear, tail, and body talk," and how to properly distinguish between a howl, a whimper, and a squeal. This chapter explains the ever-growing field of cognitive ethology — what it is, how it developed, and what it tries to do — and it ends with an evocative taste of what ethological fieldwork entails. This brief excursion is all preparation for meeting animals up close and personal.

Cognitive Ethology: Definitions and Goals

Cognitive ethology is the comparative, evolutionary, and ecological study of animal minds. It focuses on how animals think and what they feel, and this includes their emotions, beliefs, reasoning, information processing, consciousness, sentience, and self-awareness. Cognitive ethologists are interested in several things: They hope to trace mental continuity among different species; they want to discover how and why intellectual skills and emotions evolve; and they want to unlock the worlds — the umwelts — of the animals themselves. Cognitive ethologists prefer to study animals in their natural environments — or in conditions that are as close as possible to their natural environments. In today's human-dominated world, this has become more difficult to do, and an important part of many studies is to learn how the animals adapt to new conditions when their preferred and natural habitats are overtaken by trespassing humans or on occasion other nonhumans.

Interest in the emotional and mental capacities of animals has been around a long time, but the modern era of cognitive ethology, with its concentration on the evolution and evolutionary continuity of animal cognition and emotions, began with the appearance of Donald R. Griffin's 1976 book *The Question of Animal Awareness: Evolutionary Continuity of Mental Experience*. Griffin is often called "the father of cognitive ethology," and his major concern was to learn more about animal consciousness. He, too, wanted to come to terms with the difficult question of what it is like to be a particular animal. That is, he wanted to understand animals from "the inside out."

Cognitive ethology attracts a lot of attention from researchers in many different fields, including those interested in animal well-being and animal protection. I see cognitive ethology as the unifying science for understanding the cognitive, subjective, emotional, empathic, and moral lives of animals. However, when it comes to some of the challenging and exciting "big questions" that cognitive ethology raises (such as the evolution of social morality, which I address in chapter 4), the final answers require a cross-disciplinary approach. We need ethologists, geneticists, evolutionary biologists, neurobiologists, psychologists, anthropologists, philosophers, theologians, religious scholars, and religious leaders all to rise to the extremely difficult task of understanding animals' emotional and moral lives and figuring out how they compare to — and have played a role in the evolution of — human moral, ethical, and spiritual understandings.

Before we get there, however, we need to agree on some basic questions, such as, "How do we know animals think or feel anything?"

We do know this, and one reason is because of behavioral flexibility, a point stressed by Griffin. Animals display flexibility in their behavior patterns as they adapt to changing social and ecological conditions, and this shows that they are conscious and passionate and not merely "programmed" by genetic instinct to do "this" in one situation and "that" in another situation. For example, monkeys will choose not to engage in an experiment if they think they'll fail. Research has shown that rats often take a moment to reflect on what they've learned when running a maze; they pause and play back the route in their heads in reverse order and edit their experiences.

Wild animals also display behavioral flexibility depending on what's happening around them. Coyotes do well living alone, as mated pairs, or as members of packs that resemble wolf packs. How they live is strongly influenced by how much food is available — when there's enough food to feed a group, it's common to see more than two coyotes living together. Other animals also form large groups, and birds form what Ashley Ward calls "clusterflocks." A wide variety of nonhumans also show this kind of flexibility, and that is why, for example, when people speak of coyotes, wolves, and dogs, it's important not to refer to "the coyote," "the wolf," or "the dog." There is no universal member of these or any species. Of course, one of the most obvious examples is domestic dogs — including our canine companions — who are the most variable mammal around because we've bred them to be. This variability does not always serve dogs, however, since we have chosen to manufacture some breeds who have trouble breathing and who can't mate or give birth on their own.

When animals need to make decisions that involve selectively attending to specific stimuli and purposefully choosing among alternative actions, many are quite able to do so — they're aware of their surroundings and intentionally make appropriate, purposeful, and flexible choices in a wide variety of situations. Dairy cows live highly constrained, abused lives, but when given a choice — or allowed the agency — to have access to a pasture, they will push on a weighted gate to get to the grass, and they work hardest for access to the outdoors in the evening.

Flexibility in behavior is one of the litmus tests for consciousness, for a mind at work. Consciousness evolved because it allowed individuals to make choices when confronted with varying and unpredictable situations. However, once we've established that other animals have conscious minds, then we get to the really interesting questions. What are they thinking about? What do they know? What do they feel? The excitement to answer these questions is what drives cognitive ethologists. It's what gets us out of bed in the wee hours of the morning and keeps us going for long and challenging days.

Charles Darwin

Charles Darwin is often credited with being the first scientist to give serious attention to the study of animal emotions. In his book *Created from Animals*, James Rachels paraphrases Darwin as noting that mammals "experience (to greater or lesser degrees) anxiety, grief, dejection, despair, joy, love, 'tender feelings,' devotion, ill-temper, sulkiness, determination, hatred, anger, disdain, contempt, disgust, guilt, pride, helplessness, patience, surprise, astonishment, fear, horror, shame, shyness, and modesty." In his attempts to answer questions concerning the origin of emotional expression, Darwin also appears to be the first person to apply the comparative method to the study of behavior. The comparative method refers to the study of closely related and more distantly related species to learn about similarities and differences in their behavior. Darwin used six methods to study emotional expression. These were (1) observations of infants; (2) observations of the insane who, when compared to normal adults, were less able to hide their emotions; (3) judgments of facial expressions created by electrical stimulation of facial muscles; (4) analyses of paintings and sculptures; (5) cross-cultural comparisons of expressions and gestures, especially of people distant from Europeans; and (6) observations of animal expressions, especially those of domestic dogs.

In *The Expression of the Emotions in Man and Animals*, Darwin wrote about his dog with great insight: "I formerly possessed a large dog, who, like every other dog, was much pleased to go out walking. He showed his pleasure by trotting gravely before me with high steps, head much raised, moderately erected ears, and tail carried aloft but not stiffly." When Darwin would change his walking route and the dog was unsure of whether his walk would continue, "His look of dejection was known to every member of the family.... This consisted of the head drooping much, the whole body sinking a little and remaining motionless; the ears and tail falling suddenly down, but the tail was by no means wagged.... His aspect was that of piteous, hopeless dejection."

Darwin's observations of dogs and other animals ring true of much of my current research on the behavior of dogs, and I always suggest that people read his books because they are full of important information and observations that remain prescient today.

Darwin argued that emotions evolved in both animals and humans for the purpose of furthering social bonds in group-living animals. He believed that emotions connected us with the rest of our community and with the rest of the earth. My cashing out emotions as a form of "social glue" came from his and others' ideas about how shared emotions are so very important for forming and maintaining strong social bonds and groups. Furthermore, Darwin used observations of behavior such as pausing before solving problems to support his claim that even animals without language are able to reason. One of my joys is to go back to Darwin's books and see how "current" his keen observations remain today. This doesn't mean there isn't anything new under the sun, but rather it shows how essential it is to closely observe other animals — to see them as who they are and how they live — to learn about their lifestyles and minds. I wish I could have been the proverbial fly on the wall watching Darwin write and listening to him talk as his brain churned out idea after idea after idea.

In his careful research, Darwin repeatedly stressed that the differences among many species were differences in degree rather than kind. He argued that variations in mental abilities, for example, were differences along a continuum. So according to Darwin, there is evolutionary continuity among animals, not only in anatomical structures such as hearts, kidneys, and teeth, but in brains and their associated cognitive and emotional capacities. I often say if "they" — other animals — have something, so do we.

In other words, if we look hard enough, we can find the roots of our own intelligence and emotions in other animals. Again, this doesn't mean that humans and other animals are identical, but rather that they share enough common physical or functional traits that their capacities fall on a continuum. That's what "evolutionary continuity" refers to: The similarities and contrasts among species are nuances or shades of gray, not stark black-and-white differences.

Darwin didn't hedge in his arguments about evolutionary continuity. "There is no fundamental difference between man and the higher animals in their mental faculties," Darwin wrote. He attributed cognitive capacities to many animals on the basis of observation rather than controlled experiments. On occasion, lab research can preclude animals from doing everything they actually are capable of because researchers don't provide an adequate environment for them to perform natural behaviors.

In addition to making a strong case for mental continuity between humans and animals, Darwin also attributed feelings and emotions to a variety of nonhumans. For example, he claimed, "The lower animals, like man, manifestly feel pleasure and pain, happiness, and misery." Darwin also observed that monkeys were capable of deceit. About the first orangutan he saw at the London Zoo, he wrote he could see their passion and rage, sulkiness, and the very actions of despair. He observed that chimpanzees sulk and pout like children who feel disappointment, and that a rhinoceros at the zoo was "kicking and rearing out of joy."

As I describe later, current cutting-edge research agrees with Darwin's observations and ideas. We now know that dogs and other animals share with humans some of the same brain structures and some of the same neurochemicals that form the basis for such emotions as joy. Neuroimaging shows that when dogs are feeling jealous, the same area of their brain (the amygdala) lights up as it does in our brains. And the so-called "love hormone" oxytocin produced by the hypothalamus can be found in the brains of numerous nonhumans. The ability to love doesn't define us as humans, but rather it is something we share with other animals.

This process of identifying the evolutionary continuity of emotions is essentially the same as for purely physical attributes. We know that there are two-, three-, and four-chambered organs we call hearts because they pump blood. Just because the heart of a frog doesn't look like the heart of an eagle or of a human doesn't mean that a frog doesn't have a heart. And just because dog joy and chimpanzee joy and human joy aren't exactly the same doesn't mean that these animals don't experience a similar exhilaration.

Curious Naturalists

Dutch ethologist Niko Tinbergen, often called "the curious naturalist," has provided ethologists with a framework for studying animal behavior, and his suggestions are very helpful to researchers interested in cognitive ethology. In 1973, Tinbergen, Konrad Lorenz, and Karl von Frisch won the Nobel Prize in Physiology or Medicine for their pioneering work in animal behavior. Tinbergen identified four overlapping areas of research that lead to different questions about a given pattern of behavior, be it how gazelles run away from lions or how and why dogs play. He suggested that researchers should be interested in (1) the *evolution* of behavior; (2) *adaptation*, or how the performance of a specific action allows an individual to fit into their environment and ultimately allows the animal to breed; (3) *causation*, or what causes a particular behavior to occur; and (4) *ontogeny*, or development, how a behavior arises and unfolds over the course of an individual's life.

For example, if I'm interested in how and why dogs play, I'd try to answer the following four questions: (1) What do dogs do when they play and why has play evolved? (2) How does playing allow a dog to adapt to their environment, and how does playing influence an individual's reproductive fitness, or output? (3) What causes dogs to play? (4) Finally, how does play develop as individuals get older? I would follow the same line of reasoning if I wanted to know more about grief, pain, or any emotion: I'd ask why they have evolved and what they are good for. These evolutionary and developmental inquiries are what drive the study of animal minds and emotions, though in the 1990s, psychologist Gordon Burghardt added a fifth area of interest he called private experience. Tinbergen was of the opinion that subjective states couldn't really be studied with any accuracy, but he didn't deny that nonhumans experienced them.

The research methods for answering questions in each of these areas vary, but all begin with careful observation and description of the behavior patterns performed by the animals being studied. The information provided by initial observations allows a researcher to use the animal's normal behavioral repertoire to answer questions about the evolution, function, causation, and development of the behavior patterns that are exhibited in various situations. In some instances, questions about an individual's emotional state can also be pursued.

Many people were unsure how to observe and measure behavior because it "just happens and disappears," but Konrad Lorenz, also a curious naturalist, stressed that behavior is something that an animal "has" as well as what they "do." It can be thought of in the same way in which we think of an anatomical structure or organ on which natural selection can act. With careful study, we can describe an action just as we would a heart or a stomach; we can measure the action and learn why animals perform certain behavior patterns in different situations.

For example, we can ask: What do dogs do when they want to play? How do they insure that play's the name of the game and that dogs around them understand that they are playing? (Hint: They follow the "golden rules" of fair play.) And what do they feel when they play? A play-soliciting action I've studied for years is called the "play bow" or simply the "bow." With careful study, we've been able to measure the duration of the bow and how rigid and stereotyped it is. When an animal bows, they crouch on the forepaws, elevate the hind end, and may wag the tail. Once you've seen a bow, it's likely you'll recognize it again. Dogs sure do, and they know it means playtime! The bow is often used immediately before and immediately after an action that might be misinterpreted and disrupt ongoing social play (I talk more about play behavior in chapter 4). By filming the bow and watching video frames over and over, we can measure the bow, just like we can measure heart rate or the length of an arm, to find out how many seconds bows last and what they look like. This information is very important for understanding why the bow evolved as a clear and unambiguous signal that's used to tell other animals *I want to play with you.*

Tinbergen, Lorenz, and von Frisch used these methods to study phenomena such as imprinting by geese, homing by wasps, hunting by foxes, and dancing by bees. Not only that, they had fun doing so. Lorenz went so far as to don a fox coat and hop along the ground to see how geese would respond to him! Lorenz clearly loved his animal friends, and his groundbreaking work showed that by carefully studying the behavior patterns performed by different organisms (called "comparative research"), we can learn about their evolution and how their performance helped individuals survive.

Lorenz once wrote, "A greylag goose that has lost its partner shows all the symptoms that [developmental psychologist] John Bowlby has described in young human children in his famous book *Infant Grief.*...The eyes sink deep into their sockets, and the individual has an overall drooping experience, literally letting the head hang." I've seen grieving geese while cycling north of Boulder, and in December 2022, the UK news shared the story of a goose named Gretel who died of heartbreak after her "soulmate" Hansel got stuck and died in a frozen lake. I can only imagine the depth of her grief.

The Importance of Analogy

However, there is a significant difference between studying hearts and studying "heart" — the heart is a physical object, while emotions and thoughts are "invisible." All we can "see" of an emotion are the signs of it, or how it manifests itself in an animal's behavior or actions and how it affects an animal's neurochemistry. But rage, love, joy, and grief are not in themselves things that you can hold in your hand, even if you empathize with an animal and feel them in your heart.

Ethologists often make their arguments using analogies. They compare humans and other animals and look for similarities (and differences) in any number of features, including brain structure, hormones, physiology, anatomy, and genetics, as well as behavior, facial expressions, vocalizations, and so on. They look at parallels across different species and among different individuals of the same species. We are arguing by analogy when we claim that "humans have emotions that can be tied to certain brain structures, and because animals also have the same or similar brain structures, they're experiencing similar emotional states." Indeed, the brains of many species show similar neural organization in some areas involved with emotions. These arguments are thus reasonable because of evolutionary continuity among diverse animal species, including humans. Certainly, there have to be evolutionary precursors to human emotions in other animals unless we believe that human emotions came onto the scene in the absence of any animal ancestors who had emotional lives.

Here is an example of this analogous evolutionary sleuthing in action. In the 1990s, scientist Michel Cabanac discovered that reptiles, such as iguanas, maximize sensory pleasure. He found that iguanas prefer to stay warm rather than venture out into the cold to get food, whereas amphibians, such as frogs, don't show such behavior. Neither do fishes. Iguanas experience what is called "emotional fever" (a rise in body temperature) and tachycardia (increased heart rate), physiological responses that are associated with pleasure in other vertebrates, including humans.

Cabanac postulated that the first mental event to emerge into consciousness was the ability of an individual to experience the sensations of pleasure or displeasure. Cabanac's research suggests that reptiles experience basic emotional states, and that the ability to have an emotional life emerged between the time of amphibians and early reptiles. In fact, subsequent research clearly shows that reptiles have rich emotional lives, not unlike mammals, and the emotional capacities of these sentient beings shouldn't be underestimated. Female snakes have clitorises, so perhaps they enjoy having sex.

Our current understanding of reptilian behavior and emotions is well-summarized in the landmark 2021 book *The Secret Social Lives of Reptiles*. In an interview, one of the coauthors, reptile expert Gordon Burghardt, noted that he has been writing about the bias against reptiles for more than fifty years with little change. However, Burghardt is encouraged because the scientific community is now very engaged, as many of the most charismatic and large reptile species are highly endangered, such as many sea and freshwater turtles, tortoises, crocodilians, Caribbean iguanas, and even some snakes and small lizards. In May 2023, Panama granted legal rights to sea turtles "to live and have free passage in a healthy environment." The return of alligators in the US Southeast is a success story, and many people in the Southwest are now protective of their diverse rattlesnake populations. Horrible snake roundups are increasingly being targeted for elimination.

There still is a lot of work to do. Another important recent book about reptiles is *Health and Welfare of Captive Reptiles*, which brings to light the suffering of countless reptiles held in captivity. Even if the ability to have an emotional life did not first evolve among reptiles, they have one, and so they should not be treated as lesser beings than dogs, dolphins, and humans.

Fieldwork: Getting Down and Dirty

As a rule, ethologists prefer fieldwork to studying animals in a lab or a zoo. The reason for this is simple: There's simply no substitute for watching animals closely in their natural environment and collecting information on the nitty-gritty details of how they spend their time. If I begin my research, as I often do, with a deceptively simple question like, "What is it like to be a dog in such-and-such situation?" then I must try to understand how dogs get through their days and nights from their dog-centric view of the world. On many occasions I've walked around on all fours, done play bows (even on national TV), howled, barked, bitten their scruffs, and rolled over on my back — though I draw the line at mimicking the all-important mounting, humping, and hindquarter sniff (I gladly leave that for the dogs). I try to go where animals live to observe them, and as I study them, I also try to empathize with them. How would I feel if I were in the same situation? Of course, I always remember that my view of their world is not necessarily their view of their world, but the closer I can get to their view, even by personal analogy, the better I might be able to understand it.

Dogs don't typically live in a lab. In fact, very few animals do, except perhaps for those unfortunate individuals who are bred specifically for lab work. Labs can be useful as controlled environments in which to conduct research on how animal minds work, but if you really want to know how animals live, think, and feel from their point of view, then you need to join them in their world. Outside. For dogs, this could be when they're free to run on trails and at dog parks, where I spend an incredible amount of time, some say far too much. But unless I can watch free-running dogs who truly live on their own, trails and dog parks are good alternatives where I've learned tons about the emotional lives of these amazing canids.

From an ethologist's perspective, it's important to conduct research in conditions that are as close as possible to the natural environments in which natural selection occurred or is occurring. But also, in itself, the study of captive animals raises some difficult questions, ranging from the ethics of performing research on caged animals to questions about the validity of research on stressed individuals who are kept in impoverished conditions. I return to these concerns in chapter 6.

Ultimately, a good ethologist must develop an awareness of all the senses that animals use, in what combinations, and in which situations. The mix of information from different modalities is called a "composite signal." It is highly unlikely that individuals of any other species sense the world in the same way we do, and it is unlikely that even members of the same species sense the world identically, so it's important to remain alert to the possibility of individual variation. Humans are primarily a visual species, but when considering other animals, we need to factor in sounds, odors, and tastes along with visual stimuli. Also, cognitive ethologists emphasize broad comparisons among different species and don't focus on a few select representatives of limited groups of animals, such as rats, mice, or pigeons. Their general commitment to fieldwork rather than the laboratory and their interests in a broad array of animals differentiate cognitive ethologists from comparative psychologists, who are also interested in animal behavior but don't usually take a broad evolutionary or ecological perspective.

Pee, Poop, Hair, and Gas: A Day in the Life of an Ethologist

Fieldwork is about as varied as ethologists themselves. Sometimes it entails sitting back and watching, and sometimes it involves indirect or direct interactions with animals. When my students and I studied coyotes living in Grand Teton National Park, we sat on Blacktail Butte near Moose, Wyoming, for hours on end using high-powered spotting scopes and binoculars to observe known individuals. But just as often we tracked coyotes on foot, snowshoes, or skis in the cold and dismal dead of winter until we, but not they, were exhausted. When I studied Adélie penguins at Cape Crozier in Antarctica, we ran alongside them as they left their nests and went in and out of the water, and we also sat and froze as they took care of their young and stole rocks from one another in order to build nests. Fieldwork is always interesting, often exhausting, and sometimes not for the squeamish or faint of heart. To end this chapter, I offer two stories of fieldwork involving "yellow snow" and animal dung.

Back at my home in Boulder, I learned about self-awareness in dogs by using an unusual but highly appropriate research technique: collecting "yellow snow." Hair, urine, and feces contain a lot of information about who individuals are, who they're related to, what they've eaten, their hormonal and reproductive status, and their emotional state, especially whether or not they're stressed. Scientists can figure these things out using chemical analyses, but animals just use their noses. Indeed, many other animals live in very different sensory worlds from humans. They hear things we can't and smell things we're glad we can't. They inhale deeply and exhale and snort vigorously, sorting a symphony of odors through keenly evolved noses that provide them with important information and great pleasure.

Years ago I did a study to see what my dog Jethro might know about himself, specifically: Could he tell his urine from that of other dogs? I decided to move Jethro's and other dogs' "yellow snow" from one place to another over the course of five winters. I'd scoop it up in my gloves and set it down in different places so that Jethro would discover it as he cruised on the Boulder bike path where we walked and talked every day. I made sure that he didn't see me do this so he wouldn't know that I was responsible for producing the symphony of odors he was encountering. When people on the path saw me moving yellow snow, some asked me what in the world I was doing. When I told them, they nodded and walked off shaking their heads as if thinking, *Crazy academics!* Others clearly avoided me the next time they saw me. One even reported me to the local police because they thought I was nuts.

Regardless, I discovered that Jethro showed much more interest in the urine of other dogs than his own. I measured Jethro's interest by the amount of time he spent sniffing the yellow snow and whether or not he sniffed and then peed or "marked" over it. He sniffed and marked over others' urine more, and more so over the urine of other males than the urine of females, even when the samples were moved. Jethro clearly knew which urine was his and which belonged to others. Most dog owners already know that dogs discriminate their own urine from that of others, but it turns out that this hadn't before been shown experimentally outside of the laboratory. A simple noninvasive technique confirmed what we knew to be true, and subsequent controlled experiments on lab dogs have confirmed my general findings.

Collecting yellow snow in Colorado is one thing, but what I faced in July 2005, in Kenya, was another. I had the distinct pleasure of collecting elephant dung with George Wittemyer who was conducting research with Iain Douglas-Hamilton at the Samburu National Reserve in northern Kenya. Perhaps only an adventurous ethologist would respond to the offer, "Hey, you want to help me collect elephant dung?" with an excited yes. I jumped at the opportunity to accompany George, figuring at the very least it would make for great cocktail party conversation when we returned to the States.

We weren't disappointed. A few days before we went dung collecting, as I sat in a truck, a six-year-old female elephant ran toward me, stopped just short of the door, whacked the front of the truck with her trunk, and casually walked away. Then a few hours later, Hewa, a large female elephant who was a member of the group called the Wints, sauntered up to the research vehicle, looked at me as if to say, *Who do you think you are?*, and passed wind about two feet from my face. After Hewa's warm welcome, I turned to George and asked, "What's happening?"

"Oh, they're just showing you who's boss in a nice way," he told us. George said that he once had an elephant whack his vehicle with a stick and then flip the stick in the air and walk away. Elephants can flip a truck over on its side if they wish, and this once happened to George and another researcher after a bull male lost a fight. All these stories played through my mind as we sat in the truck a few days later, surrounded by a towering herd of elephants, and waited for them to go to the bathroom.

Now, elephants don't just go on our demand, so we sat and sat and sat, and as we waited, George explained these majestic creatures to us: their names, social and genetic relationships, and behaviors. It was fascinating and awe-inspiring, and best of all, we were conducting research in a noninvasive way that didn't affect the animals' normal behavior. George could have gotten much of the same information by darting and anesthetizing one of the elephants, then drawing and analyzing their blood, but how much better for the elephants (and us) to take advantage of their "natural business." Finally, one male decided it was time to go. And did he go! When he was done, George bravely ran in, disappeared among the towering elephants (who didn't seem to notice him at all), scooped the huge pile of feces into a plastic bag, ran back to the truck, and threw the bag in the rear. He was smiling all the way. I was, too, but I also couldn't wait to get rid of the bag back at the camp. This sample of dung and others like it were then sent off for genetic analyses, which helped George and his colleagues further understand the elephants at Samburu.

Elephants are among the poster species for animal emotions, since they display so many so deeply. Emotions have evolved in many species, so let's explore the passionate lives of animals.

{ 3 }

Beastly Passions: What Animals Feel

Much of chimpanzees' nonverbal communication is similar to ours. When greeting after an absence, they may kiss, embrace, or pat each other on the back. In aggressive incidents, they may swagger, scowl, scream, punch, slap, or kick. There are strong, affectionate bonds between individuals, particularly mothers and offspring, and maternal siblings, that may persist throughout life.... They show emotions clearly similar to those we label happy, sad, angry, and depressed.

— JANE GOODALL AND RAY GREEK, "THE SAD LOT OF LAB CHIMPS"

At a talk I was giving, someone once asked me if I knew of a good field guide to animal emotions, and I told them to ask their mail carrier or delivery person. Years ago when I lived in the mountains outside of Boulder, my UPS deliveryman, Dave, could read dogs better than most people I knew, and he would tell you with no hesitation that dogs experience a host of deep feelings, which they expressed to him daily. Dave once told me that one of the dogs who was a regular visitor to my house likely wasn't feeling well because she seemed "off." She was on edge and not her usual happy self. He was right; it turned out Flow had serious giardia, likely from eating too much deer poop.

Animals feel a wide range of emotions, including each of Darwin's six universal emotions: anger, happiness, sadness, disgust, fear, and surprise. If we can observe some of these emotions better than others, this probably has more to do with our inability to read the subtlety of certain feelings than with the expressiveness of animals. For when animals are happy, they feel true Happiness with a capital H, and when sad, they experience Grief with a capital G. Indeed, when we pay close attention, animals display a mindful presence, unfiltered emotions, and a zest for life.

In this chapter, I look at a variety of emotions in animals across different species. Some emotions are obvious and some are subtle; some are more-nuanced variations. We witness animals loudly sharing their joy, anger, grief, and love, but we may struggle to discern jealousy, guilt, embarrassment, and awe. Though it's rarely the focus of scientific research, animals also make jokes, display a sense of humor, and "show off." As delivery people will attest, this list of emotions is by no means exhaustive. In fact, there's every reason to believe that many animals feel just as many emotions as humans do.

That said, it's important to remember that there are differences among species in how they express their emotions (as well as in what they feel), and that there are also differences among individuals of the same species. Not all dogs or chimpanzees experience and express joy, grief, or jealousy in the same ways. Research by Sam Gosling and his colleagues has shown that each individual animal has their own personality. Individual animals can be bold, shy, playful, aggressive, sociable, curious, emotionally stable, or agreeable; they can be extroverted, introverted, dominant, or submissive. These differences are self-evident among humans, but they apply to dogs, chickens, bees, and all species. Individual and species differences make the study of animal emotions more difficult and challenging, but they also make it more exciting. As the saying goes, it takes all kinds of people to make the world go round, and the same is true within the social world of animals.

It's also important to take environmental factors into consideration. In chapter 1, I mention that rats who are infected with toxoplasmosis become "suicidal" — they develop an unhealthy attraction to cats. Who would have thought that this parasite could also influence the emotions, personality, and behavior of wild wolves? As it turns out, studies of the wolves in Yellowstone National Park have discovered that wolves infected with *Toxoplasma gondii* take more risks than those who aren't, are more aggressive, are more than eleven times more likely to leave their pack, and are more than forty-six times more likely to become pack leaders. These data show that we need to be aware that hidden factors can influence animal behavior in ways we'd never expect. Kira Cassidy, a wildlife biologist at the Yellowstone Wolf Project, who was part of the study on wolves, notes, "I think people are just starting to really appreciate that personality differences in animals are a major consideration in behavior.... Now we add a parasite-impacting behavior to the list." I couldn't agree more. Research has shown that toxoplasmosis affects humans, too.

Seeing Feelings: A Short Field Guide to Observing Emotions

Interestingly enough, it is sometimes easier to see and understand emotions in animals than in humans because animals do not filter their emotions. What they feel is clearly written on their faces, made public by gait, tails, ears, and odors, and displayed by their actions. Animal emotions are raw and on display for all to see, hear, smell, and feel. Anyone can tell if they pay close attention. For some, like my UPS delivery person, it's even an occupational necessity.

The main caveat is that identifying emotions is different from understanding the social behavior of animals. When dogs play, it's easy to see their joy, but it takes training, experience, and research to correctly interpret the complex interactions and behaviors involved in the expression of that joy, such as the "play bow" I describe in chapter 2.

In social species like wolves, where a pack needs to run like a well-oiled engine, we usually see more nuanced emotions — individuals need to know not only what others are thinking, doing, or planning but what they're feeling. If we compare highly social wolves to less-social coyotes and dogs, we find that wolves have more varied facial expressions, and that they use these expressions to communicate their emotional states to others. Wolves also use more tail positions than do dogs or coyotes to express their emotions.

Gender can also impact the way social animals express emotion. Among wild elephants, older females run the show in many social contexts. This doesn't necessarily indicate that females are more emotional than males, but we shouldn't expect, when studying social behavior, for males and females to always express emotions in the same ways, and expression might vary depending on the role of specific individuals. Females are also very important in influencing the behavior of individuals in wolf packs. Historically, the importance of females, and paying attention to gender-based differences, has often been overlooked. In her book *Bitch: On the Female of the Species*, zoologist Lucy Cooke examines how our understanding of the animal kingdom has been warped by sexist bias, starting with Charles Darwin, and how, in the past few decades, a revolution has redefined the female of the species and the very forces that shape evolution.

Nevertheless, it is surprisingly easy to recognize basic or primary emotions in animals. All we have to do is look, listen, and smell. Their faces, their eyes, and the ways in which they carry themselves can be used to make strong inferences about what they are feeling. Changes in muscle tone, posture, gait, facial expression, eye size and gaze, vocalizations, and odors (pheromones), singly and together, all indicate emotional responses to certain situations. A person doesn't even need to be consciously aware of these aspects; from simply watching an animal, people can often intuitively sense the correct emotion.

This isn't just my hunch or based on my own experience of people. Researchers have studied this and come to the same conclusion: Whether people are observing wolves, dogs, or cats, they discern emotions nearly as well as trained researchers. This means that either animal emotions are just not that well-hidden or humans have a natural ability to discern emotions in other species. I'm willing to bet it's a little of both. Try it yourself the next time you go to the park. What do you think those dogs are feeling? Here are a few things to pay attention to.

Love Is in the Air

Another way that animals communicate how they feel is through their scent. Scientists may not be very adept at using their noses, but not so other animals. Scent can be a very powerful communicator, as is made clear in the following vivid description of a male elephant experiencing what is called "musth":

> He is a hot-blooded, 30-year-old male in peak physical condition. He has mucus oozing from his cheeks and green urine streaming down his legs. His penis has a green sheen to it and he gives off a smell that can be picked up half a mile away. He wafts his ears back and forth and makes a low rumble. He looks confident: after all, many females find him irresistible. Sounds familiar? Hopefully not. He is a male elephant in musth — something like a state of rut. Sexually mature bull elephants go through musth for a one- to two-month period every year. They don't exactly hide it, excreting a cocktail of chemicals from a bulbous gland on their cheeks that can swell to the size of a basketball, passing more than 300 liters of urine a day (equivalent to 24 buckets), and — not surprisingly — smelling like a herd of goats. What's more, during this dramatic

advertisement of his sexuality the male appears to undergo something of a personality change; indeed, the word *musth* is derived from a Persian word meaning drunk. They become very aggressive and obsessed with sex, probably as a result of their high testosterone levels, which can increase by up to 60 times.

A male elephant in musth isn't an animal with whom you want to cross paths. What's interesting here is not only the rapid and obvious personality changes during musth, but also that a male in musth can communicate his intentions clearly and openly to females in whom he's interested, as well as put other males on alert. "Musth is the elephant version of expensive aftershave and a flash car. It is thought to inform males and females alike of an elephant's age, status, and reproductive health, and also increases a male's chances of reproductive success."

The chemical that does it all is called frontalin, which is secreted by sweat glands in the elephant's cheeks and also shows up in the animal's breath and urine. A male proclaims his intentions and prowess, females assess his reproductive fitness, and other males judge how strong he is before picking a fight. The precision of this signaling seems to be unique among mammals, but it's likely that nonmammals including insects also use odors to show their intentions.

About Face

Faces are also extremely important in assessing animal feelings. Charles Darwin and later researchers have stressed the importance of facial expressions in our understanding of others' emotions. For instance, it's been determined that Leonardo da Vinci's *Mona Lisa* is actually happy. A computer program using emotion recognition software at the University of Amsterdam used the curvature of the lips and the crinkles around her eyes to make this determination. She may even have been pregnant or had just given birth. While I can't say I agree, faces are nonetheless important for inferring what others are feeling and for predicting what they're likely to do in the future. Researchers have found that mammals share a great number of facial expressions, which makes it possible to infer what a monkey is feeling in the same way we can we see what an actress in a silent movie is expressing. We clearly can read the facial expressions of dogs and their wild relatives — submissive grimaces, toothy growls, open-mouth play pants — in a wide variety of social situations.

Some animals — like reptiles, fish, and birds — lack expressive faces. This can make it harder for us to interpret their feelings, but this doesn't mean that they're not feeling anything. Compelling data stemming from behavioral and neurobiological studies show that fish are conscious, intelligent, and sentient beings who express preferences. They're not merely streams of dietary proteins. Nor do birds need lips or eyebrows to be expressive with their emotions.

Look into My Eyes

The elephants came at all times of the year....Sometimes they turned their giant rumps to our thatched roof and scratched their thick hides on the rough grass, closing their eyes in what appeared to be the most blissful glee.
— Delia Owens, Secrets of the Savanna

But most disturbing of all, in Blue's large brown eyes was a new look, more painful than the look of despair: the look of disgust with human beings, with life; the look of hatred.
— Alice Walker, "Am I Blue?"

Of course, when we speak of faces, many times what we are really talking about are the eyes. Eyes are magnificently complex organs that provide a window into an individual's emotional world. As in humans, the eyes of many species reflect their feelings, whether wide open in glee or sunken in despair. Eyes are mysterious, evocative, and immediate communicators.

Doug Smith, who led the Yellowstone wolf reintroduction project for many years, once wrote about looking into the eyes of a wolf named Five: "The last time I looked into Five's eyes...she was walking away from an elk her pack had killed.... As we flew overhead, she looked up at us, as she always did. But the look she gave me had changed. To gaze into the eyes of a wild wolf is one of the holiest of grails for lovers of nature; some say what you see is untamed, unspoiled wildness.... That day in January, something had gone out of Five's eyes; she looked worried. Always before her gaze had been defiant."

Reading emotions in eyes is not as clear-cut as reading gestures like the play bow. Personal interpretation or intuition plays a role, and yet there is no more direct animal-to-animal communication than staring deeply into another's eyes. Even when we can't measure their meaning, it is the eyes that most evocatively convey sentience. Charles Siebert, in a 2006 *New York Times*

Magazine essay, wrote about the "unsettling black eyes" of Achilles, a giant Pacific octopus. "It was those eyes more than anything that I had asked [Roland] Anderson for special permission to come back and stare into on my own. Just me and Achilles."

An animal's gaze can be unflinching, and when an animal is in need, their eyes sometimes tell us all we need to know. For instance, eyes played a central role in the well-known story of Rick Swope and JoJo. While standing outside the chimpanzee enclosure at the Detroit Zoo, Rick watched as JoJo, an older 130-pound adult male chimpanzee, ran into the enclosure's moat to escape from an aggressive young male. As JoJo began to drown, Rick jumped in to save him. Three other adult male chimpanzees then began to threaten Rick, and he had to get out, but he returned moments later and succeeded in rescuing JoJo. Rick did this despite repeated warnings that his life was in danger, and when asked why, he answered: "I looked into his eyes. It was like looking into the eyes of a man. And the message was: *Won't* anybody *help me?*" In 2005, three men near my hometown of Boulder tried to save a young mountain lion who'd been hit by a car. They told reporters that the lion's eyes begged them to do so — saying the same thing: *Help me*.

Similarly, a study looking at fear in humans confirmed the importance of eye contact in recognizing emotions. It turns out that the eyes are of paramount importance in knowing that another human is feeling fear, and that people tend to look at the eyes to know when a face is fearful. The study looked at a woman who had damage to a region of her brain (the amygdala) that caused her to fail to recognize fearful facial expressions. They found the reason she couldn't perceive fear was because she didn't look spontaneously toward the eyes. Therefore, she judged a fearful face as having a neutral expression. As these stories about animals make clear, the eyes are important in perceiving fear in animals as well, and it's quite likely that the eyes are important in perceiving other emotions.

A Wink and a Nuzzle: Expressions of Gratitude

Identifying emotions with scientific rigor can be tricky, and to do so, researchers usually consider the total sum of actions, behaviors, and expressions, as well as the larger context of the situation. Even then, gut feelings can come into play. For instance, can a whale say thanks? What do you think about the following story?

In December 2005, a fifty-foot, fifty-ton female humpback whale got tangled up in crab lines, the weight of which was making it difficult for her to keep her blowhole above the water. A courageous team of divers freed her, and after being freed, the whale nuzzled each of her saviors in turn and flapped around in what one whale expert said was "a rare and remarkable encounter." James Moskito, one of the rescuers, immediately recognized that the whale was in trouble: "My heart sank when I saw all the lines wrapped around it." As for what happened afterward, he recalled, "It felt to me like it was thanking us, knowing it was free and that we had helped it." He said the whale "stopped about a foot away from me, pushed me around a little bit and had some fun." And during the rescue, he said, "When I was cutting the line going through the mouth, its eye was there winking at me, watching me.... It was an epic moment of my life."

One of the other rescuers, Mick Menigoz, was also deeply touched by the encounter: "You hate to anthropomorphize too much, but the whale was doing little dives and the guys were rubbing shoulders with it.... I don't know for sure what it was thinking, but it's something I will always remember. It was just too cool."

In this life-changing encounter, the divers expressed feeling deep sympathy and empathy for the female whale's suffering, and they describe actions by the whale that, in this context, could very well be reactions to such kindness: leviathan expressions of gratitude. We also see that it's very natural to attribute "human" feelings to animals. Indeed, the only language the divers had to describe the whale's actions was from the field of human emotion. And yet, even if our human terms are imperfect descriptions of the whale's actual emotions, does that mean the whale wasn't feeling anything? Is there another credible explanation?

In fact, another explanation appeared in print not long after this encounter. In a follow-up story in *Reader's Digest*, Frances Gulland, a veterinarian at the Marine Mammal Center in Sausalito, California, was quoted as saying that the whale probably swam in circles because her body had been kinked for so long. The divers just happened to be there while she was exercising.

Since not a single rescuer felt that the whale was "exercising," we must decide for ourselves whether the rescuers were simply projecting feelings of gratitude on an unfeeling animal, or whether they intuitively understood the whale's unfamiliar emotional gestures. I address this question of anthropomorphizing — what it is and whether it's bad — in much more detail in chapter 5. In addition, we can look for similar incidents that might support or validate any particular anecdote. While there have been others, one such occasion was in April 2023, when a whale repeatedly approached a boat captain and "asked" him to remove some lice from her body.

In the end, what we're learning is that hard science is confirming what our intuitions so often tell us: Animals express emotions in ways we are naturally able to understand.

Joy with a Capital J

Much research has been done on negative emotions — boredom, pain, fear, anger — but attention is increasingly being given to positive emotions (such as joy and pleasure), which, like some negative emotions, are rooted in areas of the ancient limbic system shared among many different mammals. Paul Ekman, a leader in the study of human emotion and facial expressions, agrees with Darwin that "the pursuit of enjoyment is a primary emotion in our lives." If this is true for us, it should also be true for other animals.

To observe animals for any length of time is to see that animals clearly enjoy themselves. Animals experience immense joy in a wide range of situations: when they play, greet friends, groom one another, are freed from confinement, sing, and perhaps even when watching others having fun. Joy is so contagious, it's essentially an epidemic. One researcher told of watching a female chimpanzee give birth, after which her closest friend screamed and embraced two other chimps. The friend then tended the mother and her offspring for several weeks.

Joy and happiness are clearly signaled by behavior — animals are relaxed and walk loosely, as if their arms and legs are attached to their bodies by rubber bands. They also speak in their own tongues — purring, barking, or squealing in contentment. Dolphins chuckle when they are happy. Greeting ceremonies in African wild dogs involve cacophonies of squealing, propeller-like tail wagging, and bounding gaits. When coyotes or wolves reunite, they gallop toward one another whining and smiling, their tails wagging wildly. Upon meeting, they lick one another's muzzles, roll over, and flail their legs. When elephants reunite, there is a raucous celebration. They flap their ears, spin about, and emit a "greeting rumble." If this behavior does not signal unashamed jubilation, then what is it — more exercising?

According to Rosamund Young in her book *The Secret Life of Cows*, even hens love to play, and they're smart, moody, emotional, and form close friendships. Although it's easy to see that chickens suffer from pain and discomfort caused by various methods of industrial farming — foot lesions, bone breakage, beak trimming, feather pecking, cannibalism, joint problems — it's more difficult to know if they're happy. But people who work with these amazing birds say they enjoy playing and hanging out with their friends. It's entirely reasonable to posit that they, like other animals, enjoy playing, and perhaps so, too, do bumblebees. Young notes that cows also play games with one another and form strong lifelong friendships. She says they also sulk, hold grudges, and act vain.

Dr. Karen Davis, president of United Poultry Concerns, calls attention to how attitudes toward certain species are driven by human self-interest, and this is particularly true with how businesses portray farm animals. She writes, "The poultry industry represents chickens bred for food as mentally vacuous, eviscerated organisms. Hens bred for commercial egg production are said to be suited to a caged environment, with no need for personal space or normal foraging and social activity. They are characterized as aggressive cannibals who, notwithstanding their otherwise mindless passivity and affinity for cages, cannot live together in a cage without first having a portion of their sensitive beaks burned off — otherwise, it is said, they will tear each other up." However, Davis continues, "My experience with chickens for more than twenty years has shown me that chickens are conscious and emotional beings with adaptable sociability and a range of intentions and personalities. If there is one trait above all that leaps to my mind in thinking about chickens when they are enjoying their lives and pursuing their own interests, it is cheerfulness."

As the ethologist, animal advocate, and author Jonathan Balcombe points out, evolution favors sensory rewards because it drives beings to stay alive and reproduce. We prefer what gives us pleasure, so evolution has made what we need to do pleasurable. To quote Balcombe:

> For too long scientists have denied the existence of positive sensory experiences in other species because we cannot know for certain what another being feels. But in the absence of compelling evidence to the contrary, it is more reasonable to assume that other creatures, who share so much in common with us through our shared evolutionary origins, do, in fact, experience pleasure. We cannot feel the hummingbird's response to a trumpet-flower's nectar, the dog's anticipation of chasing a ball, or the turtle's experience of basking in the sun, but we can imagine those feelings based on our own experiences of similar situations. What we can observe in animals, combined with our capacity to empathize from our own experience, leaves little doubt that the animal kingdom is a rich repository of pleasure. And as we grow to accept and acknowledge

the pleasure that attends animals' lives, evidence for it will proliferate, for we are more likely to find something when we are looking for it.

Evidence of joy in animals is so extensive that it should hardly need further discussion. At a meeting on animal sentience that I attended in 2005, John Webster, a professor of animal husbandry at the University of Bristol, said, "Sentient animals have the capacity to experience pleasure and are motivated to seek it.... You only have to watch how cows and lambs both seek and enjoy pleasure when they lie with their heads raised to the sun on a perfect English summer's day. Just like humans."

Jonathan Balcombe also provides a nice example:

During a trip to Assateague, Virginia, I watched two fish crows (*Corvus ossifragus*) land on an old wooden billboard that protruded incongruously from a cattail marsh. Hoping they would stay awhile, I swiveled my telescope and focused on them. They first engaged in flight play, then over the next ten minutes, one bird (always the same one) repeatedly sidled up to the other, leaned over, and pointed his/her beak down, exposing the nape. The other bird responded by gently sweeping his/her bill through the feathers as though searching for parasites. There was every indication that they were mates or good buddies, and that their contact was as pleasurable for both giver and receiver as a massage or caress between two humans.

Joy in animals may be obvious to one's eyes, but it is also confirmed by neurobiological studies, or "hard science," on the effects of play and laughter.

These findings suggest that there are neurochemical bases for why play is enjoyable, and that the same chemical changes occur in both animals and humans during play. In other words, a boy and his dog wrestling in the yard are not only both playing — they both understand that they're playing, and they're getting the same pleasurable feelings from doing so.

Having a Laugh

Laughter clearly isn't uniquely human. There's solid scientific information showing that dogs laugh — there's what researcher Patricia Simonet called "a breathy, pronounced, forced exhalation" heard when dogs are excited and when they play. Simonet discovered that the sounds of dog laughter could also soothe other dogs who hear them even if they're not playing. Rats also chirp with joy, and there's little doubt that continuing research will show this to be true in many other animals.

Jane Goodall noted that female chimpanzees occasionally use sticks to tickle their genitals and laugh as they're doing it. Laughing, or vocalizations such as play pants, might be responsible for the reciprocity of social play. Primatologist Takahisa Matsusaka has shown that play panting during social play in wild chimpanzees encourages continuation of tickling or chasing and reduces the risk of play escalating into aggression. And Robert Provine suggests that laughter actually evolved from labored breathing due to physical play, and it likely signals *I like it, do it again.*

Neural circuits for laughter can be found in very ancient regions of the brain. As neuroscientist Jaak Panksepp notes, "Research on rough-housing play in mammals, both sapient and otherwise, clearly indicates that the sources of play and laughter in the brain are both instinctual and subcortical." The neurochemical dopamine is also implicated in both human and rat laughter.

Panksepp says, "Tickles are the key.... They open up a previously hidden world." Rats who are tickled bond to the researchers and seek out tickles. Their feelings function as social glue. Rats laugh only when they're feeling good, in the same sense that animals play only when all is well.

"It's like the behavior of young children," says Robert Provine. "A tickle and laughter are the first means of communication between a mother and her baby, so laughter appears by about four months after birth."

In humans, the importance of such an early behavior is apparent, so it shouldn't be a surprise to find that it's equally important among other animals. There's nothing frivolous about the study of animal laughter. But do animals know any good jokes?

A Sense of Humor: Stand-Up Animal Comedians

It's one thing to discover the neurochemicals responsible for the feelings of joy and happiness in animals, but how do we distinguish and correctly identify subtler, more deliberate thoughts and emotions? If animals can laugh, does that mean they can tell jokes? Many pet owners will nod affirmatively: They know for themselves that animals have what we call a sense of humor. Farmers know to be wary around their donkeys. My late canine companion Jethro used to run all over with his favorite toy, a stuffed rabbit, in his mouth. He'd scoop it up on the run, bound here and there, and then turn to see if anyone was watching. If they were, he'd continue his frantic game. My friends and I would always laugh, which apparently was the reward Jethro was looking for.

Other anecdotal evidence seems to confirm that animals not only know how to have fun but they know how to make fun, tell jokes, and engage in slapstick comedy. The following two stories were told to me by friends.

Jill Robinson — the founder of Animals Asia who rescued Jasper — shared with me further stories about the moon bears she has rescued. She wrote:

> The moon bears will often gang up together against another bear — in fact, we have one female group of bears we affectionately call the "knitting circle" because they remind us so much of elderly ladies who seemingly have nothing better to do but gossip the day away and who will join together to warn away another bear who comes too close to their exclusive circle. Such friendships for a species recognized to be solitary in the wild are extraordinary. Often, one bear will wait until another is distracted and then steal toys or food away — in every sense like a greedy sibling, biding their time until the opportunity arises to steal from a brother or sister. Their sense of fun and mischief finally wins through all but the most severe legacy of pain and stress.

Some years ago renowned filmmaker and writer Michael Tobias told me about his family's large scarlet macaw, Mac, who Michael and his wife, Jane Gray Morrison, took care of for decades. Actually, Michael said they served this magnificent and rare emperor of an avian on their proverbial hands and knees, and he claimed that Mac apparently wielded more verb tenses than most highly literate *Homo sapiens* and was endowed with a macaw vocabulary that defied casual analysis. He also, it seems, was endowed with an infectious sense of humor and sport.

Mac giggled, chuckled, and roared with laughter; he teased all who came near, squealed with delight when running around the house, bounded up and down on trees outside, and even played "magic carpet" — wherein his human slaves raced down hallways dragging large towels with the macaw riding aboard. Michael has observed laughter in numerous parrot species — as well as in a host of other birds — but he believes parrots may be among the greatest wits on the planet. Not only do they laugh, but they display perfect timing. If Mac lobbed a small Frisbee at Michael with his enormous beak and hit Michael in the face, or if Michael missed it, the macaw nearly fell over in hysterics. With a specific knowledge of where people are ticklish, the macaw then charged Michael's underarm so as to goose him until Michael himself was in tears laughing.

And what does Michael make of all this? He wrote:

> If a bird, whose kind are headed to extinction given current rates of human trespass and insanity; a bird that will outlive most individual humans; one whose intelligence far surpasses most ungainly, carnivorous, indifferent *Homo sapiens*; can nonetheless find humor in the madness of the world, share joy with the few people it trusts, then we must take heed before the awesome innocence that the rest of creation is endeavoring to offer up by way of an example. If we fail to appreciate, even worship the laughter of macaws, or to engage in the dialogue that the animal kingdom is extending to us; or, finally, to be quiet, humble, and serene before the sheer miracle of life, then we will indeed go down in biological history as the worst, meanest, shortest-lived failure in the history of evolution. We would be wise indeed to take the lesson of a forgiving macaw to heart, and none too soon.

Seeking Attention, Showing Off, and Pride:
Hey, Look at Me!

A number of animals — particularly dogs, cats, and apes — have been observed "showing off" in deliberate ways to attract attention or to proudly display either themselves or some object they possess. Whether or not other animals display pride is still hotly debated — it's hard to know what animals feel when they engage in creative, playful, attention-seeking behavior. My own "educated guess," and that of others, is that a sense of pride could be part of some displays, but that isn't necessary, and it's not necessarily the only emotion. "Showing off" might be part of game playing, joking around, having a sense of humor, or even simply sharing something pleasurable. As important as the emotions involved, when an animal is "showing off," it suggests that they have a theory of mind, since the behavior implies: *If I do this, I know what you'll think or feel.*

In 2022, a news story entitled "Cool Leaf! Study Records Chimp Showing Off Object in Human-Like Way" described this behavior in chimpanzees:

"Chimpanzees show each other objects just for the sake of it, researchers have found, revealing it isn't only humans who like to draw attention to items that have captured their interest." Summarizing an observation of an adult female chimpanzee, Fiona, showing her mother, Sutherland, a leaf, the researchers wrote, "Our observations suggest that in highly specific social conditions, wild chimpanzees, like humans, may use referential showing gestures to direct others' attention to objects simply for the sake of sharing. The difference between humans and our closest living relatives in this regard may be quantitative rather than qualitative, with ramifications for our understanding of the evolution of human social cognition."

While I find this anecdote fascinating, I think the conclusions of the researchers display "ape exceptionalism." Surely, primates, human and nonhuman, are not the only species who like to show off. For example, Deb Pilley Bianchi told me an interesting story about Chaser, the famous dog who knew a thousand words. Chaser was a trickster and enjoyed the attention she received when displaying her amazing intellect for others. In a March 2023 email, Bianchi wrote:

> The only time Chaser had the zoomies was when we were in a TV studio during dress rehearsals. She was literally a "dog gone wild," and I think it's because she was cooped up in my Brooklyn home. One time in particular, she had disappeared in the NBC studio while we were talking to the producer and was barking somewhere off set. She never barked, so I was wondering if there was another dog on the set. She appeared out of the prop room with a beach ball, balancing and bouncing it on her nose while propelling it forward, never touching the ground. The stagehands were amazed, asking me how we taught her to do that. We never did teach that to her, and I never saw her do it again, but she sure was having fun.

Clearly, Chaser was bored and letting off some pent-up steam, but she also made sure everyone else saw her perform her new "trick."

I once asked fifty people with dogs, "Do you think your dog shows off?" All said yes, forty-seven emphatically, and five people who also had cats said they thought their cats did, too. I have personally witnessed many dogs picking up a stick or a ball and using it to get other dogs to chase them. In these cases, a desirable object is used to inspire others to engage in play. For instance, my dog Jethro and his canine friends — Zeke, Maddy, Zoe, and

Lolita — had free rein to run around near my mountain home when I was there (so could monitor what they were up to). I used to enrich their daily lives by burying small bits of food in the grassy fields around my house, and they would spend upward of an hour looking for the treats before settling down. On a number of occasions, if one of the dogs found a morsel and the others were looking elsewhere or resting, the dog would parade in front of the others, step away, and look back to see if they were now looking. Once the others were paying attention, the dog who found food would try to get one or more of them to follow or chase, and often play and zoomies ensued. On a number of occasions, cached food was shared and they then went about their own business.

In my long-term study of wild coyotes, I saw youngsters and adults doing what I and my field assistants called "showing off," in the same manner as dogs. Coyotes would parade around with a bone or part of a carcass in their mouth, and when others noticed and paid attention, the coyote would run away or, on some occasions, share the food.

I also received a very interesting story from Neva Davis about her cat Obi:

> With dogs, I find it hard to tell if they are showing off when they prance around holding a toy, or if they are trying to entice the other dogs to chase them to initiate play. But I do see this behavior of getting the toy (or other desired object, even a stick) and prancing around with it in front of the other dogs. I contrast this to times when they really want to keep something to themselves. In those cases, they usually grab it and slink away from the other dogs, so they won't have to share.
>
> However, one incident with a cat struck me as blatant showing off. This cat has since passed away, but his name was Obi. This was during a time when we had about six cats in the house.... A cicada got inside and all the cats, even the elderly ones, gave chase. Every single cat wanted to catch the cicada. Obi caught the cicada and then pranced around, up and down the halls, showing every other cat that he had the cicada. He showed off to us, making sure we saw, and then dashed off before we could take the bug away from him. I don't think he was trying to initiate play, since the cats never really play fought over an object, and he did this routine even with cats he didn't really get along with or were so old they never played with him.

Awe, Wonder, and Altered Emotional States: Chimpanzees and Waterfalls

If animals can display such humorous delight in other beings, is it also possible that they might, at times, marvel at their surroundings and feel a sense of awe at the world? Do they experience the joy of simply being alive? And if so, how would that express itself? Wild animals spend upward of 90 percent of their time resting: What are they thinking and feeling as they gaze about? It would be nice to know. Again, science may never be able to measure such emotions with any precision, but anecdotal evidence and careful observation indicate feelings akin to awe and wonder may exist among nonhumans.

Sometimes animals "just go nuts," as one of my students once commented. Indeed, when we were at Lake Nakuru in Kenya, in July 2005, we watched a young black rhinoceros "go nuts" — he ran all over the place like he was having a fit, while his mother watched him with a careful eye all the time. What was causing him to act with such abandon? We don't know. There was no cause we could see other than that it felt good. There were no other potential playmates around, or for that matter no other rhinos except for his mother.

Then there are chimpanzee waterfall dances, which are a delight to witness. Sometimes a chimpanzee, usually an adult male, will dance at a waterfall with total abandon. Why? The actions are deliberate but obscure. Could it be they are a joyous response to being alive, or even an expression of the chimp's awe of nature? Where, after all, might human spiritual impulses originate? Jane Goodall wonders whether these dances are indicative of religious behavior, precursors of religious ritual. She described a chimpanzee approaching one of these falls with slightly bristled hair, a sign of heightened arousal:

> As he gets closer, and the roar of the falling water gets louder, his pace quickens, his hair becomes fully erect, and upon reaching the stream he may perform a magnificent display close to the foot of the falls. Standing upright, he sways rhythmically from foot to foot, stamping in the shallow, rushing water, picking up and hurling great rocks. Sometimes he climbs up the slender vines that hang down from the trees high above and swings out into the spray of the falling water. This "waterfall dance" may last ten or fifteen minutes.

Chimpanzees also dance at the onset of heavy rains and during violent gusts of wind. Goodall asks, "Is it not possible that these performances are stimulated by feelings akin to wonder and awe? After a waterfall display the performer may sit on a rock, his eyes following the falling water. What is it, this water?" In June 2006, Jane and I visited the Mona Foundation's chimpanzee sanctuary near Girona, Spain. We were told that Marco, one of the rescued chimpanzees, does a dance during thunderstorms in which he looks like he's in a trance. Perhaps numerous animals engage in these rituals but we haven't been lucky enough to see them.

Goodall wonders, "If the chimpanzee could share his feelings and questions with others, might these wild elemental displays become ritualized into some form of animistic religion? Would they worship the falls, the deluge from the sky, the thunder and lightning — the gods of the elements? So all-powerful; so incomprehensible."

These are wonderful and important questions, and my observations of many different wild animals as well as companion dogs has made me ask similar ones. I've observed dogs staring into space for minutes on end, and I always wonder what they're thinking and feeling. Some years ago, a woman at a local dog park asked me similar questions and told me her dog Maxine looked mesmerized and "locked into" sunsets of a particular kind — when there was an orange sun highlighted against milky white clouds — though Maxine never showed such focus and awe for other sunsets.

In 2023, a study was published about rapid spinning in captive great apes that proposed they do this most likely to induce altered states of consciousness and generate emotional "highs." This research supports Jane's observations, since awe is certainly an altered state of consciousness, and it is most welcomed because some people doubt uncontrolled observations of wild animals. All in all, it's highly unlikely that humans are the only animal blessed with a sense of wonder.

Jane admits that she'd love to get into the minds of animals even for a few moments. I would, too. I would love to know what Maxine was thinking and feeling when she saw specific sunsets. It would be worth years of research to discover what animals see and feel when they look at the stars — what an amazing experience it would be.

Grief and Sadness

Tragedy struck in 1997. Fourteen's mate, Thirteen, was much older than she....In March, his collar emitted the accelerated series of beeps that signals a dead wolf. After that, Fourteen took off. She traveled west through

> *inhospitable terrain that she had never ventured into before....Eventually Fourteen made her way back to her family....No one, including myself, wanted to suggest that she had traveled alone so far because she was mourning the loss of her mate. But she never bred again even though she consorted with other mature males.*
>
> — Doug Smith, "Meet Five, Nine, and Fourteen, Yellowstone's Heroine Wolves"

Many years ago I realized that there is *no* question about whether animals grieve. As the chapter 1 stories about grieving birds show, the universal signs of grief are seen most keenly when animals respond to the death of a mate, family member, or friend. Like humans, animals can suffer monumentally over a separation or loss. In her book *How Animals Grieve*, Barbara King writes, "It's not anthropomorphic to use this label for them.... Grief and love are not human qualities. They are things we share with some other animals."

Grieving animals may withdraw from their group and seek seclusion, resistant to all attempts by others to draw them out. They may sit in one place, motionless, staring vacantly into space. They stop eating and lose interest in sex. Sometimes they become obsessed with the dead individual. They may try to revive them and, failing that, remain with the carcass for days on end. Janet Baker-Carr writes in *An Extravagance of Donkeys* that these wonderful beasts of burden show concern when a group member becomes ill. The author tells about the death of a female, and how the other donkeys stood on her grave and brayed a sad requiem long into the night.

We're surely not the only animals who grieve, and it's narrowly anthropocentric and arrogant to think we are. Humans and other animals share neural pathways when it comes to suffering. Some scientists even say that the demeanor of elephants suffering from the loss of friends and the disruption of social bonds resembles post-traumatic stress disorder (PTSD). Scientists have also identified a part of the brain associated with PTSD, called the ventromedial prefrontal cortex; activity in this part of the brain varies among people, making some more prone to anxiety, fear, and stress than others. In animals this structure seems to be important with helping the brain to forget. It's highly likely that domesticated, wild, and urban animals suffer from severe depression and PTSD, and it's known that abandoned dogs can suffer from PTSD because their lives can be incredibly stressful.

Grief itself is something of a mystery, for there doesn't seem to be any obvious adaptive value to it in an evolutionary sense. It does not seem to increase an individual's reproductive success. Some theorize that perhaps mourning strengthens social bonds among survivors who band together to pay their last respects. This may enhance group cohesion at a time when it's likely to be weakened. Whatever its value, grief is the price of commitment, that wellspring of both happiness and sorrow.

Consider Sadie and Oscar, two senior dogs who lived together. One night, according to family members who witnessed the dogs, Oscar seemed to be doing very well and went to bed as usual on his pillow downstairs. As usual, Sadie, Oscar's best friend, went to sleep near Oscar. Then around midnight, Oscar got up and walked to the steps that led upstairs and tried in vain to gain footing, *something he'd never done before.* As Oscar sprawled out at the foot of the steps, the noise awoke Sadie, who went over to see what was happening. She sat with Oscar for a few minutes, sniffing his body and whining softly, and then she seemed extremely upset and flew up the steps, running directly to the bed where the father of the family was sleeping. By the time he got down the steps, Oscar had died, and Sadie was moping around and seemed out of sorts. She knew something was wrong and wouldn't leave Oscar's body. Did Sadie know Oscar was dying? Was she trying to get help for Oscar by alerting her human? And immediately afterward, was she grieving her lost friend? Just what she knew and felt remains a mystery, but surely her head and heart were moved in powerful ways.

A Fox Funeral

One evening when I was driving up my road, I saw a large tan animal walking toward my car. I thought it was my neighbor Robb's German shepherd, Lolo, so I stopped and opened the car door to say hello. As I came face to face with a big male mountain lion, only then did I hear Lolo barking behind me. The lion stared at me, seemed to shrug his shoulders, and walked off, as if to say, *Silly human.* I was so scared, I slammed the door shut, went home, and ran straight to my house.

The next morning Robb told me that Lolo had found the carcass of a red fox, and I went to look at it. The fox, a formerly very healthy male, had obviously been killed by the lion, and his body was partially covered with branches, dirt, and some of the fox's own fur.

Two mornings later, I waited until it was fully light and headed out to hike with Jethro. (No more surprises for me!) I looked down the road and saw a small female red fox trying to cover the carcass. I was fascinated, for she was deliberately orienting her body so that when she kicked debris with her hind legs, it would cover her friend, perhaps her mate. A family of foxes had lived near my house for almost a decade, and it was very likely she was related to, or at least a close friend of, the dead fox. She'd kick dirt, stop, look at the carcass, and intentionally kick again. I observed this "ritual" for about a minute. Afterward, the female continued to slink around the carcass with her tail down. A few hours later, I returned to the carcass and found it totally buried.

Had I just seen a fox funeral? It looked to me as if the female fox was trying to bury the dead fox. Her actions and her manner certainly reflected sadness and grief. I was lucky to see what I did. Much happens in the complex lives of animals that we can't see — and will never be able to re-create in a lab — and when we're fortunate to observe animals at work, it is a splendid event that can reveal much about their innermost feelings.

Deadly Crossings

Family-living prairie dogs who many people write off as "merely rodents" also grieve. A few years ago while riding my bike north of Boulder, I observed a moving interaction between an adult black-tailed prairie dog, who looked to be a female, and a youngster, most likely her child, who had been killed by a car. It looked like the accident had occurred a few minutes before I happened on the sorrowful scene, and I stopped and dictated some notes into my phone to document what I saw:

Five times, the adult prairie dog tried to retrieve the carcass of the smaller prairie dog off the road. She clearly was trying to remove the carcass from the road, and eventually, the cars stopped and allowed her to finish. She dragged the carcass about ten feet away, looked at me and looked at the carcass, and then went back to the carcass and touched it lightly with her forepaws. After this, she walked away emitting a very high-pitched vocalization. I waited a few minutes to see what else she would do, and as she moved back toward the carcass again, she looked at me and stopped — so I left because I didn't want to disrupt her saying goodbye, if that was what she was doing. Minutes later, I caught up with another rider who was ahead of me, and he told me he saw the prairie dog trying to remove the carcass from the road twice.

I described this to prairie dog expert Dr. Con Slobodchikoff, who replied in an email:

> Amazing! But I am not surprised. I haven't seen this with prairie dogs, but I recently saw a male quail trying to push a dead female

off the road. The female had been killed by a car and was lying in the middle of the road. The male kept nudging her and pulling her, oblivious to all the cars that were going right past him. Like you, I left because I didn't want to disturb him. Sadly, when I went back, I found that he had managed to pull the female halfway across the road, and then some insensitive and perhaps distracted motorist ran over him and killed him as well. I think that all sentient beings grieve for their mates, their relatives, their friends. We're just too blind to see it.

Given what I saw and what Slobodchikoff wrote, I was especially pleased to receive an email from Anet Barnhill in July 2023, who wrote:

Today I witnessed the same thing you wrote about while driving the Alpine Loop here in Alpine, Utah. A small prairie dog had been hit in the road and a bit larger prairie dog was trying with all its might to pull it off the road. I wondered what was going on, so I've been searching the web and found your story. Clearly I was very moved by what I saw…so sad for these two friends or family. I hope they were able to get off the side of the road before anyone else drove past after me. It's stayed in my mind since this morning, and I wanted to know what the behavior meant. Thank you for sharing your story.

I love hearing from people — citizen scientists I often call them — who let me know about all sorts of animal behavior that clearly indicates emotions and sentience.

The Comfort of Friends

Primates, being near relatives to humans, often exhibit behavior remarkably close to us. Baboons are known to seek the comfort of friends after deaths in their family. Researchers from the University of Pennsylvania report that baboons rely on friendships to help them cope with stressful situations. In their study, when a lion killed a baboon named Sierra, her mother, Sylvia, looked to friends for support. Said researcher Anne Engh, "With Sierra gone, Sylvia experienced what could only really be described as depression, corresponding with an increase in her glucocorticoid levels."

When baboons experience stress they show an increase in hormones called glucocorticoids, just as humans do. These hormones are produced by the adrenal gland. Baboons can lower their glucocorticoid levels through friendly social contact, such as by expanding their social network after the loss of specific close companions. "Without Sierra, Sylvia really had nobody else," Engh said. "So great was her need for social bonding that Sylvia began grooming with a female of a much lower status, behavior that would otherwise be beneath her."

Engh concluded somewhat cautiously: "Our findings do not necessarily suggest that baboons experience grief like humans do, but they do offer evidence of the importance of social bonds amongst baboons. Like humans, baboons seem to rely on friendly relationships to help them cope with stressful situations."

Dying of a Broken Heart

Many years ago, veterinarian Marty Becker gave his father a miniature schnauzer, Pepsi, as a gift. Marty told me that he had helped to deliver Pepsi, who was the runt of his litter, and the dog became his father's best friend. For years they shared the same food, the same chair, and the same bed. Then, when he was eighty years old, Marty's father committed suicide. Soon after family, friends, and the police left his house, Pepsi ran downstairs to the spot in the basement where Marty's father had died and stood as rigid as a statue. When Marty picked Pepsi up, the dog went from rigid to limp in his arms and emitted a painful moan. Marty put him in his father's bed and Pepsi immediately fell asleep. Marty later found out from his mother that Pepsi hadn't been in the basement for ten years because he was afraid of steps. Had Pepsi overcome his fear in order to say goodbye to his lifelong friend? Pepsi never recovered from his companion's death. Remaining weak and withdrawn, he slowly died. When Marty buried Pepsi, he was convinced that Pepsi had indeed died of a broken heart: He lost the will to live once the human to whom he was so closely bonded and devoted was no longer around.

An Elephant Burial

Elephants are well known for the deep concern and curiosity they show for dead individuals. It's rare to witness elephants coping with their dead, so it's difficult to collect detailed data, but numerous observations show time and again that elephants are concerned with suffering and death and openly show compassion when encountering another elephant who is in distress or when discovering a corpse. Here, elephant expert Cynthia Moss describes one such encounter in her book *Elephant Memories*:

> They stood around Tina's carcass, touching it gently.... Because it was rocky and the ground was wet, there was no loose dirt; but they tried to dig into it... and when they managed to get a little earth up they sprinkled it over the body. Trista, Tia, and some of the others went off and broke branches from the surrounding low brushes and brought them back and placed them on the carcass.... By nightfall they had nearly buried her with branches and earth. Then they stood vigil over her for most of the night and only as dawn was approaching did they reluctantly begin to walk away.

Iain Douglas-Hamilton and his colleagues have shown that elephants also extend this compassion to nonrelatives — to those who aren't genetically related. And elephants display a well-documented fascination with death and bones. Many experts describe seeing these magnificent mammoths become excited and agitated when they come across a dead elephant. Karen McComb, an expert on animal communication and cognition, and her colleagues performed a unique field experiment to study the concern that elephants show for the dead. They presented skulls and other objects to nineteen groups of wild elephants. They discovered that the animals preferred to investigate elephant bones and tusks and could even distinguish elephant skulls from those of other species. Researchers also found that the wild elephants spent twice as much time examining the elephant skulls as they did the buffalo and rhino skulls.

And what, besides love and despair, could possibly be animating the elephants in the following story? Here, Cynthia Moss describes the actions of the same elephant family above after a group member had been shot:

> Teresia and Trista became frantic and knelt down and tried to lift her up. They worked their tusks under her back and under her head. At one point they succeeded in lifting her into a sitting position but her body flopped back down. Her family tried everything to rouse her, kicking and tusking her, and Tullulah even went off and collected a trunkful of grass and tried to stuff it in her mouth.

Tahlequah: A Very Tragic Tour of Grief

In 2018, a mother orca, twenty-year-old Tahlequah, a member of an endangered population of killer whales near San Juan Island, Washington, was observed grieving the loss of her child. For seventeen days, she was seen balancing the corpse of one of her children on her head and on her nose, and other pod members also took turns in what was called a funeral ritual. This event became a "hot" global media item. Jenny Atkinson, director of the Whale Museum on San Juan Island, noted, "Ceremonies can go on for days to honor and mourn the loss of a loved one.... I think that what you're seeing is the depth of importance of this calf and the grief of the mother and the family." Dr. Ken Balcomb, a pioneering whale researcher, orca expert, and founder of the Center for Whale Research, called it "a very tragic tour of grief." I agree. If Tahlequah wasn't exhibiting undying love and grieving along with other pod members, what were they doing?

Further, why did this become such a media sensation? I think it's because many people don't identify with orcas or other marine animals as they do with, for example, companion animals and terrestrial mammals. If Tahlequah had been a dog, chimpanzee, gorilla, or elephant, there wouldn't have been such media hype. But orcas don't have expressive tails, ears, faces, gaits, or postures, and so we don't know what cues to look for to discern what they are feeling. However, Tahlequah's behavior was so remarkable and persistent, we couldn't ignore it or mistake it: Clearly, she was expressing strong feelings, and the closest human analogy is grief. In 2020, Tahlequah gave birth to a son, Phoenix, and perhaps this has helped her overcome her enduring sorrow.

Speaking about whales, in December 2022, Port Townsend, Washington, recognized the rights of endangered southern resident orcas. Orcas are highly sentient beings who care deeply about what happens to themselves and their family members and friends.

Eternally Devoted: Llamas Teach a Lesson in Grieving

I'd like to end this section on grief by sharing a story that my friend Betsy Webb told to me. Betsy lives in Homer, Alaska, and has lived with llamas for years. Her moving story is an elegant testament to the sentient awareness and emotional depth of animals. Betsy wrote:

> Llamas are gregarious by nature, extremely perceptive, and forge deep bonds with one another. In the pasture, our llamas often feed in the same area, sleep next to each other, and stay close together

when they face off an unfamiliar animal or predator. On the trail, they become extremely agitated if they lose sight of each other when one stops to rest and falls behind. They vocalize quite a bit. My favorite is their delicate greeting call, which sounds like a miniature bagpipe exhaling.

When my family moved from Colorado to Alaska, we brought our two Colorado llamas with us. As fate would have it, we inherited two Alaska llamas with our new house and grounds. Each twosome had spent their lives together. At first, the twosomes were a bit standoffish, but in time, they became fast friends and a foursome.

Several years later, the oldest llama, Boone, died quite suddenly at twenty-seven years old. One day, he laid down on his side, too weak to get up. The next day, his life partner, Bridger, died in the same fashion, next to him. It was early spring and the ground was still frozen, so we hired a friend with a backhoe to prepare their grave just across the fence. We carefully hoisted Boone and Bridger over the fence and into the ground, and then covered them. The other pair, Taffy and Pumpernickel, stood by and watched the entire process quietly.

For the next two days, stoic Taffy stood across the fence from the grave and stared at the hole in the ground. She barely moved from the spot. Excitable Pumpernickel stayed in his little barn and wailed for two days. On the third day, they emerged from their grieving and resumed their normal activities.

Did Bridger surrender himself to death following the loss of his lifelong buddy Boone? And Taffy and Pumpernickel, both very distinct personalities, grieved in their own personal ways. For me, the most moving memory of losing two llamas so close together was experiencing the caring and harmonious llama death and grieving process.

Love: Where Science and Poetry Meet

Love. Has there ever been a more troubling and mysterious emotion? Humans have struggled to understand and define love since the dawn of consciousness, so what possible hope is there that we can understand and define love in animals? And yet, though we don't truly understand love, we do not deny its existence, nor do we deny its power. We experience or witness love every day, in a hundred different forms; as I note above, grief is the price of love.

As with grief, there is no doubt that animals love. One reason we know is because the powerful "love hormone" oxytocin is widespread among mammals, while other animals have their own form of oxytocin — birds and reptiles have mesotocin, and fishes have isotocin. Of course, the word is often misused. People can be quick to assume love or exaggerate it, such as the false claim that dogs "love unconditionally." They don't. In fact, dogs are quite selective about whom they choose to love.

Love is perhaps the most complex of all emotions, given its bewildering variety of forms and shadings. On this landscape, where science and poetry meet, we find love that is romantic, parental, filial, and erotic, and we see love express itself as friendship, loyalty, affection, tenderness, devotion, commitment, and compassion.

If I were to hazard a definition, one that we could use to examine animal behavior in search of love, I'd say that love means preferring the close company of another individual, seeking them out when they're not around, and if necessary protecting and caring for them. It means forming and maintaining strong and close reciprocal social bonds and communicating feelings of close attachment between loved ones. Not exactly poetry, but it's a start.

There's considerable evidence that many animals are capable of feelings that run the gamut of the varieties of love. The brain machinery of love — the neuroanatomy and neurochemistry that allow us to feel love — is similar or identical to that of numerous other animals. Once again, science is finally catching up with what our intuitions already tell us, and in the following sections, I look at love that is romantic (that involves selecting and keeping a mate), maternal (that involves parent-child bonding), and filial (that involves love between siblings or friends).

Love and Marriage

So much research has been done on the mating styles and habits of animals that it's clear that romantic love isn't uniquely human, nor, I should add, are humans role models for marriage or committed love. How much of this is chemical and how much of this is purely "emotional" is a matter of debate, even regarding human love. But anyone who's lived with a cat or a dog — or watched cows, sheep, or pigs — knows that animals fall in love. And as with all their emotions, their love is as pure and unfiltered as it can be. There's no evidence that nonhumans play around with love and try to make their spouses jealous. While individuals of some species do have love affairs and engage in intentional or opportunistic sneak copulations, there's no evidence that they flaunt it.

In many species, individuals take the time to get to know one another and to reinforce their close bonds. Romantic love can be slow to develop, and mating rituals can take a lot of time and energy. These rituals even sometimes involve physical risk, and sometimes vows need to be renewed. Male marmosets spend a good deal of time figuring out who to mate with. Evidence from functional magnetic resonance imaging (fMRI) studies shows that they make careful decisions and evaluate what they're getting into before they mate. In many canids — that is, members of the dog family, which includes wolves, coyotes, foxes, jackals, and dingoes — males and females who have mated for years still greet one another like they're long-lost friends and court vigorously, even though they've done it before.

And just to be clear, "courtship" is not a euphemism for sex; it describes the same romantic dance that you would witness at any high school prom. For example, zoologist Bernd Würsig describes the courtship of southern right whales, which he observed off Peninsula Valdis, Argentina. While courting, the male and female whales continuously touch flippers, beginning with a slow caressing motion. Then they roll toward each other, briefly locking both sets of flippers as in a hug, and roll back up, lying side by side. They then swim off, side by side, touching, surfacing, and diving in unison. Würsig says he followed a pair for about an hour, during which they continued to travel together.

Whales are fellow mammals, but there's compelling evidence that fish also make choices about mating. They're not automatons. Researcher Lee Dugatkin observed what he calls "guppy love." Dugatkin discovered that males change their behavior and become bolder in response to a predator when there is a female around because females find bold males more attractive. Even among fish, it seems like males will risk it all for love.

Helen Fisher, author of *Why We Love*, put forth what she calls an "immodest proposal" concerning the evolution of love: "All these data lead me to believe that animals big and little are biologically driven to prefer, pursue, and possess specific mating partners; there is chemistry to animal attraction. And this chemistry must be the precursor to human love."

The most devoted mates are not necessarily our closest kin, the great apes, or other mammals. More than 90 percent of bird species are monogamous, and many mate for life. Fewer mammals are monogamous, and the nonhuman primates appear comparatively callous when it comes to commitment. Chimpanzee males, for example, don't spend much time courting, mating, or remaining with a female whose young they've fathered. When males aren't needed to provide protection or food for their mates or their offspring, they frequently try to mate with as many females as possible. That sounds like "happy hour" at the watering hole to me.

A Mother's Love

Just as romantic love among animals has been well documented, so, too, has maternal love. This love is sometimes displayed most dramatically when a child is in trouble; mothers of many species will fight to the death to protect their offspring, and when a child is hurt or killed, they exhibit the deepest feelings of grief and pain. Sea lion mothers, watching their babies being eaten by orcas, squeal eerily and wail pitifully in anguish over their loss, and dolphins have been observed struggling to save a dead infant.

Mother love couldn't be more primal. In the first known example of an invertebrate caring for their young, Australian leeches, or bloodsuckers, have been shown to be devoted parents. They carry their newborns and nurture their young for up to six weeks after hatching. They also protect their children from predators and carry them to new areas where they'll be safe.

Mother love is found in innumerable species. Orcas may not be very nice to the animals they eat, but they are good, very loving parents, as Tahlequah, the orca mother, displayed when she carried the carcass of her dead child for over two weeks. Baby whales are born ready to swim, and like human children, they are very curious infants. Naomi Rose, who works at the Humane Society of the United States, once watched a large group of orcas along the coast of Vancouver Island in Canada. One day Naomi saw a mother surface without her calf and discovered that the young orca had swum behind her boat. The calf was playing and having a good time exploring his surroundings, but his head was very close to the propeller. In a few minutes, the calf's mother arrived next to the boat, but rather than chase her calf away, she simply kept a close eye on him. She tolerated his curiosity and playful spirit, while remaining watchful for danger. Could it be that this orca mother was trying to negotiate that delicate balance all mothers face: wanting to keep her child safe, but not be overprotective?

Elephants are well known to be very devoted parents, and elephant expert Cynthia Moss tells the following story of a mother's devotion. In late February of one year, Echo, the "beautiful matriarch" of her elephant family, gave birth to a male, Ely, who could not stand up because his front legs were bent. He was born with rigid carpal joints. Echo continually tried to lift Ely by reaching her trunk under and around him. Each time Ely stood, he was able to shuffle around on his knees for a short while, but then he would collapse to the ground.

Eventually, the other clan members left, and Echo and her nine-year-old daughter, Enid, stayed with Ely. Despite the fact that Echo and Enid were hungry and thirsty, they wouldn't leave an exhausted Ely to go to the watering hole. After a great deal of effort, all three elephants were able to reach the watering hole, where Echo and Enid splashed themselves as well as Ely. Echo and Enid then made low rumbling calls to the rest of their family. After three days, Ely's joints loosened, and he finally was able to stand on his own.

There is more to the story. When Ely was seven years old, he suffered a serious wound from a spear that was embedded about one foot into his back. Although Echo now had another calf, she remained strongly bonded to Ely and would not allow a team of veterinarians to tend to him. When Ely fell down after being tranquilized, Echo and other clan members tried to lift him. Echo, Enid, and another of Echo's daughters, Eliot, remained near Ely despite attempts by the veterinarians to disperse the elephants so that they could help Ely. The elephants refused to leave despite gunshots being fired over their heads. Finally, Ely was treated and survived the injury.

Love Is Blind

One could argue that attracting a mate and protecting one's offspring are genetically hardwired instincts. Is what looks like "love" actually unthinking evolution protecting its investment? There may be some truth to this — the desire to reproduce certainly seems instinctual — but biological needs alone don't explain the wide range of emotions animals display, and they certainly don't explain the stories that follow. In these, animals are seen acting with incredible devotion and care even though there's no biological advantage to be had.

Incredibly, I've heard of more than one story of a sighted animal helping a blind animal. One is the tale of Annie, a blind mule, and her constant companion and guide, Charlie, a steer. Both were permanent residents of Colorado's Black Forest Animal Sanctuary (BFAS), a nonprofit rescue organization. Annie was nursing a broken shoulder and bound for the slaughterhouse when BFAS volunteers brought her in. She lived at the ranch for more than a year before she met Charlie. Initially, Charlie and Annie were kept in separate pastures, but during one cold spell all the livestock were herded into a single pen for warmth. Charlie hit it off with Annie. He started nuzzling up to her and playing with her, and the two became inseparable. Before, Annie had a hard time finding the water tank, but Charlie unfailingly led her to it. She followed him around the pasture, avoiding bumping into the fence, and they slept cuddled up together.

There are numerous stories of animals of different species forming close social bonds that resemble what we call love. A one-year-old hippopotamus, named Owen by his caretakers, formed a close bond with a century-old male tortoise named Mzee (Swahili for "old man") after floods in Kenya (due to a 2004 tsunami) left Owen dehydrated and alone. Owen was found by wildlife rangers near the Indian Ocean and brought to a wildlife sanctuary in Mombasa. Afterward, Owen slept and ate with his tortoise friend, and the two remained inseparable, close buddies.

Love Dogs

Tika and her longtime mate, Kobuk, two wonderful malamutes, enjoyed their last years together in the home of Anne, who shared their love story with me. Kobuk was charming, energetic, and always demanded attention. He'd always let you know when he wanted his belly rubbed or his ears scratched. He also was quite vocal and howled his way into everyone's heart. Tika was quieter and pretty low-key. If anyone tried to rub Tika's ears or belly, Kobuk shoved his way in. Tika knew not to eat her food unless it was far away from Kobuk. If Tika happened to get in Kobuk's way when he headed to the door, she usually got knocked over as he charged past her.

Then one day a small lump appeared on Tika's leg. It was diagnosed as a malignant tumor. Overnight Kobuk's behavior changed. He became subdued and wouldn't leave Tika's side. Then Tika had to have her leg amputated and had trouble getting around. Kobuk, clearly worried about her, stopped shoving her aside and stopped caring if she was allowed to get on the bed without him.

About two weeks after Tika's surgery, Kobuk woke Anne in the middle of the night the way he did when he really needed to go outside. Tika was in another room, and Kobuk ran over to her. Anne got Tika up and took both dogs outside, but the dogs just lay down on the grass. Tika was whining softly, and Anne saw that Tika's belly was huge and swollen. Anne realized that Tika was going into shock, so she rushed her to the emergency animal clinic in Boulder. The veterinarian operated on her and was able to save her life.

If Kobuk hadn't fetched Anne, Tika almost certainly would have died. Tika recovered, and as her health improved after the amputation and operation, Kobuk became the bossy dog he'd always been, even as Tika walked around on three legs. But Anne had witnessed their true relationship. Kobuk and Tika, a true old married couple, would always be there for each other, even if their personalities would never change. They were love dogs doing for each other what needed to be done.

In our dealings with animals, we would be wise to follow their example. It is certainly my fervent hope that one day we will.

Embarrassment: Can a Monkey Blush?

Do animals ever feel embarrassed? It may seem like a silly question, but the ability to be embarrassed might be one sign of a sentient, self-aware being. If you have no sense of self, then it doesn't matter who's looking, right?

Embarrassment is difficult to observe; by definition it's a feeling that one tries to hide. But Jane Goodall believes she has observed what could be called embarrassment in chimpanzees. Fifi was a female chimpanzee whom Jane knew for more than forty years (she was the daughter of an equally famous female, Flo). When Fifi's oldest child, Freud, was five and a half years old, his uncle, Fifi's brother Figan, was the alpha male of their chimpanzee community. Freud always followed Figan; he hero-worshipped the big male. Once, as Fifi groomed Figan, Freud climbed up the thin stem of a wild plantain. When he reached the leafy crown, he began swaying wildly back and forth. Had he been a human child, we would have said he was showing off. Suddenly the stem broke and Freud tumbled into the long grass. He was not hurt. He landed close to Jane, and as his head emerged from the grass, she saw him look over at Figan — had he noticed? If he had, he paid no attention but went on grooming. Freud very quietly climbed another tree and began to feed.

Marc Hauser observed what could be called embarrassment in a male rhesus monkey. After mating with a female, the male strutted away and accidentally fell into a ditch. He stood up and quickly looked around. After sensing that no other monkeys had seen him tumble, he marched off, back high, head and tail up, as if nothing had happened.

A few years ago a woman named Monica shared a similar story with me about her dog Rufus who, she was convinced, had experienced embarrassment. Rufus was hot on the trail of Lori, a good-looking mutt who flirted with many other dogs at a local dog park. Rufus was a cool dog who hung out and watched other dogs courting Lori but didn't do much more. After two males left Lori's side because she didn't seem to be interested in them, Rufus began zooming around Lori, perhaps to display his athletic prowess and show off his aerobic fitness. When he was running toward Lori, he tripped. When he got up, Lori was staring at him, and Rufus stopped, looked around ostensibly to see if anyone else saw him tumble, shook off the fall as if nothing had happened, ran up to Lori, did a play bow, and tried to mount her. Lori was willing, and afterward, they played relentlessly for around ten minutes. It turned out Lori was not in heat or at least fertile — she'd been spayed — but possibly she was putting out some pheromones that the male dogs misinterpreted as her being ready to make one of them the lucky father.

Of course, three anecdotes, however intriguing, aren't proof. Once again, comparative research in neurobiology, endocrinology, and ethology is needed to learn more about the subjective nature of embarrassment. If we study the neural and hormonal correlates of embarrassment in humans and see similar patterns in animals (as we've done for joy), then we're on safe ground claiming that animals are capable of experiencing embarrassment. Yet these anecdotes raise the possibility, and there's no good reason to think animals can't.

Anger, Aggression, and Street Gangs: Red in Tooth and Claw

As with joy and fear, there's no question that animals can get angry in the same ways that humans get angry. We share common neurochemicals (such as serotonin and testosterone) and brain structures (such as the hypothalamus) that are important in the expression and feeling of anger, aggression, and revenge. It's easy to identify anger and aggression as well, for it often appears raw and unfiltered.

In her book *Bird Minds: Cognition and Behaviour of Australian Native Birds*, bird expert Gisela Kaplan writes about the behavior of cooperative white-winged choughs. She notes that these highly innovative birds live in cohesive groups, but they also "can turn into street gangs, sending out scouts to destroy magpie-lark nests and even nests of other choughs in their own group. They steal youngsters from other groups and there are helpers at the nest that pretend to help when they do not." Sounds like some mammals I know, including members of my own species!

Birds are also well known for their sibling conflicts. While siblings can be friends — they will protect one another and cooperate against a common enemy — they will also evict one another from the safety of a nest, steal food from one another (even in times of plenty), and sometimes kill one another (which is called siblicide). In American pelicans, sibling fighting is common, and fights to the death occur in black eagles, cattle egrets, great blue herons, and Mexican blue-footed boobies. Usually, the oldest and largest siblings start fights. In one study of black eagles, the older chick relentlessly pursued his younger sibling and pecked him more than fifteen hundred times during thirty-eight attacks. The senior chick gained fifty grams during this period, whereas the younger chick lost eighteen grams.

Siblicide is obviously driven by the competition to survive, but that makes it no less emotional for the combatants. There are actually fewer good examples of siblicide in mammals because much rivalry occurs in dens and cannot be seen. But it is documented in both Galapagos fur seals and spotted hyenas. In general, same-sex aggression is more common and heated than aggression between the sexes.

In spring 2023, orcas made international news for trying to sink boats off the coast of Spain in the Strait of Gibraltar and around Galicia. One story involved three orcas who worked together to sink a yacht by ramming into it. Some orca experts thought these encounters were occurring because of a traumatic interaction between a female orca and a boat — the killer whales were "pissed off" and getting revenge. Spanish ethologist David Nieto Macein agreed, and in May 2023, he emailed me the following note:

> I know these orcas personally....Indeed, they started because they approached a poaching boat, surely to test if they found tuna... but this boat severely wounded them with harpoons and gave them electric shocks....Since then, those two killer whales have attacked similar boats on the keel and length, and little by little others have been learning and imitating....Due to other injuries committed against them from fishing boats, the orcas are angry and attack certain types of boats to stop them. The more aversive they are, the angrier they are. Instead, they approach my friend's boat without a problem and friendly.

Despite Macein's on-the-ground observations, some people suggested the attacks were actually a new form of play, and the whales were just having fun and enjoying themselves. Actually, more than one explanation might be correct; like people, animals can have multiple motivations. However, even if this was partly a form of rough-and-tumble play fighting, the aggressiveness was unmistakable. My own take is that it was likely a bit of both. What's also interesting is that the orcas appeared to develop different strategies of harassing the boats, and this behavior was spreading throughout the population of orcas.

Nasty Nick: Stories of a Mean and Pissy Baboon

Nick, an olive baboon, was an adolescent when he joined what was known as the "Forest Troop" in the southeast corner of Masai Mara National Reserve in Kenya, and even then you could almost see contempt on his face. Or so says world-famous Stanford professor Robert Sapolsky, who described Nick in *A Primate's Memoir*. Sapolsky wrote that Nick dominated his age group, and "he was confident, unflinching, and played dirty." One time, Nick trounced another baboon named Reuben in a fight, and afterward Reuben "stuck his ass up in the air" — which is a sign of submission, giving up. Nick went over as if to examine his bottom and then slashed him with his canines. Sapolsky is known for speaking eloquently and plainly about animal behavior, whether it sounds like anthropomorphizing or not, and he was equally blunt about Nick:

> The guy simply wasn't nice. . . . He harassed the females, swatted at kids, bullied ancient Gums and Limp. On one memorable day, he took exception to something that poor nervous Ruth had done and chased her up a tree. Typically, at this point, the female takes advantage of one of those rare instances when it pays to be smaller than the males — she goes to the farthest end of a flimsy branch and hangs on for dear life, depending on the fact that the heavier male can't crawl out to where she is and bite her. And, typically, the male, thwarted, positions himself to at least trap the female, keeping her screaming on the precarious branch until he gets bored. So Ruth gallops up the tree, Nick after her, and Ruth leaps out to a safe edge. Nick promptly climbs onto a stronger, thicker branch directly above her. And then urinates on her head.

Nick was a bully and a competitive Darwinian nightmare, but he was also infuriatingly smart. In another episode, Sapolsky had darted Reuben with a sedative as part of an experiment in behavioral physiology. After darting, a baboon becomes unconscious, and Sapolsky and his field team can handle them, in this case to determine the relationship between the individual's dominance rank, personality, and physiology. However, as Reuben was slowly going under, Nick approached. Reuben lifted his head, saw Nick, and gave him a fear grimace. Then Nick placed his hand on Reuben's shoulder and the other on his haunch, leaned back, and bellowed a "wahoo" call. After holding this pose, Nick marched back into the underbrush. "I couldn't believe it," Sapolsky wrote. "That bastard had just taken credit for my darting."

This is not to say that anger is always a function of survival competition or plain meanness. For instance, in Saudi Arabia, a man hit and killed a baboon with his car. Afterward, the baboon's troop lay in waiting for three days by the side of the road until the same driver appeared again. As the driver passed the troop, one baboon screamed, and then all the baboons threw stones at the car and tore out its windshield.

Please, Don't Tease the Chimpanzee

Finally, here is another cautionary tale. Ron Schusterman, a marine biologist at the University of California–Santa Cruz, loves to tell the story of Franz, a young male chimpanzee who was kept in a lab and known to be a feces thrower. Ron's friend Larry was one of Franz's favorite targets. One day, Larry noticed that Franz's cage had been cleaned up, and so he teased Franz: "You can't get me — na na na na na." Franz stared at Larry while he was being taunted, and when Larry finished, Franz regurgitated some partially digested food he had been fed a few minutes earlier and threw it at Larry, splattering him across the front. Franz then ran around in a victory dance.

Are There Autistic Coyotes and Bipolar Wolves?

Because it's usually ignored, I want to pose a final question in this chapter: If animals feel many, if not most, of the emotions humans feel, can they also become mentally impaired? While we see emotions being freely and openly expressed in a wide variety of species, often there are individuals who seem to be "out of it." For example, on occasion I've seen a young animal who just doesn't seem to get it, an individual who just doesn't know how to play. I remember a coyote pup named Harry who didn't respond to play signals by playing, as did most of his littermates. Harry also didn't use play bows very often and just didn't seem to have a clue about how to initiate play, or even how to play if he got to do it at all. For a long time I simply chalked it up to individual variation, figuring that since behavior among members of the same species can vary, Harry wasn't all that surprising.

Some years ago I was asked if there were autistic animals, and I thought about Harry and realized I wasn't sure. Because there are autistic humans, there likely are nonhuman animals who suffer from what might be called autism. Perhaps Harry suffered from coyote autism. Simon Baron-Cohen has made great strides in learning about human autism using ethological studies, and ethologist Niko Tinbergen eventually turned his attention to the study of autism, so there may indeed be a useful connection.

I remember other animals. There was another coyote, a large male named Joe, who seemed to go all over the place. He'd often seem to be sulking and moping around for no obvious reason and then instantaneously run around as if he were happy, seemingly without a care in the world. Then there was Lucy, a young wolf who behaved similarly to Joe. Some days Lucy behaved "normally," like a typical wolf, whereas on others she was either really wired or really down. Other colleagues have also remarked that on occasion one of the animals they're watching seems to be very unusual. But we never thought to call the out-of-the-ordinary individuals autistic or bipolar. I already mention that nonhumans, wild and domestic, can suffer from PTSD.

Perhaps, to be consistent with arguments about evolutionary continuity and emotions, this would not be out of order. As I note above, experienced ethologists and psychologists now believe that elephants hurt like us and heal from psychological trauma like us and suffer from post-traumatic stress disorder. (They have a huge hippocampus, a brain structure in the limbic system that's important in processing emotions.) Then how far do we need to stretch to include autistic and bipolar animals? Many different psychological disorders have been diagnosed in dogs, so there's no reason why this couldn't be true for their wild relatives and other creatures.

Now let's look at an even more intriguing possibility: that animals possess moral sensibilities, and that these are the evolutionary precursors of our own moral behavior.

{ 4 }

Wild Justice, Empathy, and Fair Play: Finding Honor among Beasts

Those communities which included the greatest number of the most sympathetic members would flourish best and rear the greatest number of offspring.

— Charles Darwin,
The Descent of Man and Selection in Relation to Sex

I've long been interested in play behavior. This might sound like a frivolous field of study — a number of my colleagues certainly told me so when I first started — but after many years of examining videos of playing dogs, coyotes, and wolves and trying to understand why animals play the way they do, I have been led to ask a series of big and ultimately surprising questions: Do animals play fair? Do they negotiate agreements to play (as opposed to fighting or mating), and do those agreements require cooperating, forgiving, apologizing, and admitting when they're wrong, as well as trusting others? Are animals honest? If one breaks their agreement, do they consider that wrong? Are there consequences for doing something wrong? If animals demonstrate a dislike for getting the short end of the stick or being shortchanged, does that indicate that animals have a sense of justice, of right and wrong, good and bad — does that mean, in other words, that animals are moral beings? And if animals can be shown to display a sense of justice along with a wide range of cognitive and emotional capacities, including empathy and reciprocity, does that make the differences between humans and all other animals a matter of degree rather than kind?

Finally, if all this is true, then is morality in fact an evolved trait? Does "being fair" mean being more fit — does being more virtuous improve an individual's reproductive fitness, while being less virtuous harms it? To put it another way, do nice guys, gals, and their genes last longest? Do the friendliest and nicest survive best?

These are indeed big, complicated, difficult questions, but mounting evidence points straight to the conclusion that there is "honor among beasts." While much of the research that's been aired widely deals with nonhuman primates, there are also compelling data from studies on social carnivores that support the claim that moral behavior is more widespread among animals than previously thought.

Based on my long-term detailed studies of play in social carnivores — including wolves, coyotes, red foxes, and domestic dogs — I believe we can make the stronger claim that some animals other than nonhuman primates might be — truly are — moral beings. Other ethologists (such as Nobel laureate Niko Tinbergen and the well-known field biologist George Schaller) stress that we might learn more about the evolution of human behavior from studies of social carnivores than from studies of other primates because the social behavior and social organization of many carnivores resemble that of early hominids in a number of important ways (divisions of labor, food sharing, care of young, and intrasexual and intersexual dominance hierarchies). Given this, social carnivores may hold the key for unlocking the nuances of animal morality.

So let's get into the nitty-gritty and seriously consider how play figures into discussions of morality. To begin with, when animals play, there are rules of social engagement that must be followed, and when these break down, play suffers. Animal play appears to rely on the universal human value of the golden rule — do unto others as you would have them do unto you. Following this requires empathy (feeling another's feelings) and implies reciprocity (getting paid back for favors assuming that others follow the same rule). Further, in the social arena, animals who don't play well don't seem to do as well as those who do play. Darwin might very well have been right when he speculated that more cooperative and empathic individuals have more reproductive success — they survive better and are more likely to have their own children. At the end of this chapter, I propose that this means we should make another paradigm shift in how we understand animals and ourselves. "Survival of the fittest" has always been used to refer to the most successful competitor, but in fact cooperation may be of equal or more importance. It is likely that individual survival, for any species, requires both to some degree, while for social species (as opposed to asocial species) the balance may shift significantly, with the most cooperative individuals most often "winning" the evolutionary race.

We still need to learn more about the importance of cooperative behavior and its relationship to wild justice. Indeed, it's only in the past twenty or so years that discussing morality in animals hasn't automatically been met with a skeptical raised eyebrow and a disdainful laugh. Jessica Pierce and I write about this in detail in our 2009 book *Wild Justice: The Moral Lives of Animals*, on which we were working when the first edition of this book was published. Traditionally, morality has been the exclusive right of humans, even sometimes the very definition of our humanness, and some scientists and others still vehemently resist the idea that we might actually share this quality with other beings. Yet more and more biologists, neuroscientists, philosophers, and ethologists are beginning to think that morality may be a broadly adaptive strategy that has evolved in many species. I'm not saying animal moral behavior is the same as human moral behavior. Rather, my proposal is that the phenomenon to which "morality" refers is a wide-ranging biological necessity for social living. Just as emotions are a gift of our ancestors, so, too, are the basic ingredients of morality: namely, cooperation, empathy, fairness, justice, and trust.

Distinguishing Morals and Ethics

Before going further, it's important to make clear what I'm *not* proposing. Philosophers sometimes make a distinction between morals and ethics. Ethics is the philosophical study of moral beliefs and behaviors (equivalent to "moral philosophy"). Ethics suggests the contemplative study of subtle questions of rightness or fairness. Ethics is wondering why fairness exists, why one action is considered "fair" and another isn't. Here, I'm arguing that some animals have moral codes of behavior, but not that animals have ethics. They may sit around, paw to chin, regarding the world like Rodin's *Thinker*, but I don't think they are contemplating "why good is good." As far as we know, this is a distinctively human phenomenon.

The word *moral* was first coined in the fourteenth century as an extension of the Latin *mos*, "custom." *Mos* described the proper behavior of a person in society, and it referred especially to one's manners. In its most basic form, morality can be thought of as "prosocial" behavior — behavior aimed at promoting (or at least not diminishing) the welfare of others. Morality is an essentially *social* phenomenon: It arises in the interactions between and among individuals, and it exists as a kind of webbing or fabric that holds together a complicated tapestry of social relationships. The word *morality* has since become shorthand for knowing the difference between right and wrong, between being good and being bad.

In the context of animals, morality refers to a wide-ranging suite of social behaviors; it is an internalized set of rules for how to act within a community. Moral behavior includes (but may not be limited to) cooperation, reciprocity, empathy, and helping. Morality has emotional, or affective, components, and it also has cognitive components. Jessica Pierce sees human morality experienced as, and also enforced through, a number of regulating mechanisms: anger, indignation, guilt, conscience, shame, reputation, ostracism, retribution, joy, love, disgust, and desire. More research needs to be done on how animals enforce and regulate morality.

Defining "Fairness"

Animals exhibit fairness during play, and they react negatively to unfair behavior. In this context, "fairness" has to do with an individual's specific social expectations and not some universally defined standard of right and wrong. If I expect a friend to play with me and they act in an aggressive manner — dominating or hitting me rather than cooperating and "playing" — then I will feel I am being treated unfairly because of a lapse in social etiquette. We have found, by studying the details and dynamics of social play behavior in animals, that animals exhibit a similar sense of fairness. For instance, one way we know that animals have social expectations is that they show surprise when things don't go "right" during play, and only further communication keeps play going. For example, during play when a dog becomes too assertive, too aggressive, or tries to mate, the other dog may cock their head from side to side and squint, as if they're wondering what went "wrong." For a moment, the violation of trust stops play, and play continues only if the playmate "apologizes" by indicating through gestures, such as a play bow, their intention to keep playing.

Social play is thus based on a "foundation of fairness." Play occurs only if animals agree and consent to cooperate, if they have no other agenda but to play, and if, for the time they are playing, they put aside or neutralize any inequalities in physical size and social rank. As we will see, large and small animals can play together, and high- and low-ranking individuals can play together, but not if one of them takes advantage of their superior strength or status.

Morality in Animals: They Have It, Too

People who resist the idea of "moral animals" tend to hold one of two opposing beliefs: Either it's impossible because humans are clearly the only beings blessed with such virtues, or it's impossible because all animals, including humans, are obviously inherently immoral and amoral. I'd like to quickly address both these reactions, since they can keep us from having an open mind and seeing animals as they actually are.

Those who would put humans on a pedestal above all other creatures feel threatened by the possibility of morality in animals, since it seems to threaten the special and unique status of humans. This idea that humans are the most virtuous creatures usually comes from religion, so to say animals can be moral is sometimes perceived as a threat against some deeply held religious belief. But it's not. This isn't an either-or situation: Either humans are special or no one is special. Both can be true: Humans have ethics and spiritual awareness, and animals display moral codes of behavior. The only thing that is threatened by the proposition that animals are moral beings is the belief that morality alone defines our "human nature." To me, what sets us apart is our ability to conceive of a "human nature." As I say above, dogs are dogs, and they probably don't ponder their "dogness" — but they do know when another dog breaks their trust, and they may shame that dog or avoid them in the future. In fact, it makes good evolutionary sense to assume that morality evolved in other animals, and it wouldn't be the first time that a trait used to separate humans from other animals actually brought them closer together. For instance, before Jane Goodall's groundbreaking discovery, it was assumed that tool use was uniquely human, but now we know that many primates as well as dogs, corvids, rats, and other animals fashion and use tools.

For instance, an orangutan named Fu Manchu once used a wire to pick the lock to his enclosure at the Omaha Zoo to let himself and some of his friends out. He hid the wire between his lips and his gums, and once free, none of the orangutans tried to leave the zoo. Rather, they hung out in other places, especially by the elm trees near the elephant enclosure. Wild gorillas, similarly, have been seen dismantling poachers' traps. Shortly after a poacher's trap killed a gorilla named Ngwino in a Rwandan forest, two youngsters — Rwema, a male, and Dukore, a female — went out and worked together to destroy traps in the forest in which they lived.

Then there are the pessimists. At a social hour at a meeting of the Animal Behavior Society, one of my esteemed colleagues heard me talking about animal morality and walked away scratching his head thinking I'd had too much beer. He was especially concerned because of the abundance of immoral behavior that humans display daily. "So," he exclaimed, "how can you talk about morality in animals when we're so #&$#^$%#* amoral?"

On one level, he's right: Humans can be selfish, unfair, and uncaring, and their moral codes can sometimes be self-servingly hypocritical. Just take a cursory glance at the front page of the newspaper: The murder of a family during a robbery is considered unacceptable, but not so killing in self-defense or as part of a distant, "justified" war. Humans can lie, steal, and cheat, and they can justify their actions so they never feel "wrong." At times, it can be hard to imagine how anyone could consider humans morally "above" any other animal beings.

However, it is just as easy to find examples of human kindness, compassion, and generosity. Humans are capable of both virtuous and less-than-virtuous behavior, just as nonhuman animals show themselves capable of fair and unfair behavior. I'm not proposing that animals are necessarily more virtuous than humans, but simply that many species, to varying degrees, have social standards for behavior. Individuals may break their standards or uphold them, but they all understand them, and they understand that there are social consequences based on what they do. Neither am I proposing that dog morality is the same as chimpanzee morality, or that chimpanzee morality is the same as human morality, and so on. But I do believe we can find the evolutionary roots of what we call human morality by carefully studying the behavior of other animals.

Jerome and Ferd: Two Dogs at Play

Years ago one of my students, Josh, called me and, with much excitement in his voice, told me the following story of watching his 120-pound malamute, Jerome, engaged in play:

> I saw the most amazing thing today at Mount Sanitas. Jerome wanted to play with a strange dog named Ferd who was about a quarter his size. Jerome bowed, barked, wagged his tail, rolled

over on his back, leapt up, and bowed again, all to no avail. Ferd just stood there with surprising indifference. But about a minute later, while Jerome was sniffing a bush where a large mutt had just peed, Ferd strolled over and launched onto Jerome's neck, and bit him hard and was sort of hanging in midair, legs off the ground. I thought, this is it; Jerome will kill this little monster.

And you know what? Jerome shrugged Ferd off like a fly on his back, turned around and bowed, and then took the little guy's head into his mouth and gently mouthed him. They then played for about half an hour, during which Jerome never ever was very assertive or unfair. He'd bite Ferd softly, roll over, paw at his friend's face, and swat him lightly. Then when things got rough, and Ferd backed off with his tail down and cocking his head from side to side — trying to figure out if he was a goner — Jerome would bow again and they'd play some more. Jerome seemed to know that he had to be nice and fair in order to play with his little buddy. Ferd knew what Jerome wanted, and Jerome knew what Ferd wanted, and they worked together to get it. Man, dogs are smart. I couldn't believe it.

Josh was a good student. He understood the "language" of animal play (which I describe next), and so he was able to "read" this encounter and all of its internal communications. However, I'm willing to guess that anyone watching these two mismatched dogs would have been able to tell, after a few minutes, that they were playing and not fighting — just as we can tell at a distance whether two boys who are wrestling really mean to hurt each other or whether they are just kidding around. This is because, when animals play, they must agree to play. They must cooperate and behave fairly, and the language of cooperation is easy to recognize. Further, when cooperation and fairness break down, play not only stops, it becomes impossible. Uncooperative play is an oxymoron, and that is a large reason why play is such a clear window into the moral lives of animals.

What Is Play?

Most of my research on play has involved domestic dogs and their wild relatives, coyotes and wolves — all of whom are canids, or members of the dog family — so while I focus on the animals I know best in this chapter, there are ample examples from other animals that support my views on play and social morality. Young cats, chimpanzees, bears, and rats, to name only a few, love to play to exhaustion. When I originally wrote this chapter, I could watch two red foxes playing outside my office almost every morning. One invited the other to play by bowing, the other responded, and then they'd wrestle, rearing up on their hind legs and screaming and boxing, chasing one another, resting, and then playing some more. If one bit the other too hard, there was a brief pause during which they looked at each other to make sure all was okay — that this was still play — and then resumed romping. They were negotiating with each other to maintain the rules of fair play. And so long as they kept negotiating, they felt comfortable playing very vigorously because they shared a common goal and knew that neither would try to beat up the other.

I think of play as being characterized by what I call the "Five Ss of Play": its spirit, symmetry, synchrony, sacredness, and soulfulness. The *spirit* of play is laid bare for all to see as animals wildly run about, wrestle, and knock one another over. The *symmetry* and *synchrony* of play is reflected in the harmony of the mutual agreements to trust one another — individuals share intentions to cooperate with one another to prevent play from spilling over into fighting. This trust is *sacred*. Finally, there's deepness to play, for animals become so completely immersed in it that I like to say they are the play. Play is thus a *soulful* activity, an expression of the essence of an individual's being.

There's also incredible freedom and creativity in the flow of play. This is easy to see and amazing to watch. I refer to this as the "Six Fs of Play": its flexibility, freedom, friendship, frolic, fun, and flow. As they run about, jump on one another, somersault, and bite one another, animals re-create a mind-boggling array of scenarios and social behaviors. It's difficult to believe that when animals are deep into play they can actually keep track of what they are doing, but they can. It's possible that animals are "practicing" and "rehearsing" important behaviors that will help them to survive. As animals play, it's not unusual to see known mating behaviors intermixed in highly variable kaleidoscopic sequences along with actions that are used during fighting, looking for prey, and avoiding becoming someone else's dinner. In no other activity but play do we see all of these attributes and behaviors occurring together.

Fun, Fun, Fun: Why Animals Play

Animals love to play because play is fun, and fun is its own very powerful reward. Some people write off "fun" as being unscientific, but they are wrong. Data and a touch of common sense clearly show that fun does exist in numerous diverse nonhumans. In fact, in 2015, an entire issue of the scientific journal *Current Biology* was devoted to the evolution and biology of fun, and it covered dogs, dolphins, apes, birds, fishes, frogs, reptiles, and invertebrates. Yes, it's fun to study fun in nonhuman animals, but it's also scientifically valid to ask why and how fun has evolved. Because if something is fun, animals will do it, and in the case of play, there are many reasons why play is adaptive and has evolved in diverse species. When skeptics say something like, "We really don't know if animals are having fun," a scientifically sound response is what citizen scientist David Matheson wrote in a 2023 letter to the editor at *New Scientist*: "If it looks like fun and has any other necessary attributes of fun, chances are it is some sort of fun." Absolutely! This is how we evaluate young infants before they can speak: If children are doing things that seem like play, that evoke fun, then that tells us what they're feeling.

Dogs and other animals seek out play relentlessly, and it's very difficult to get them to stop; normal animals don't usually intentionally seek out activities they don't enjoy. The joy and fun associated with play is so strong that it outweighs the possible risks, such as injury, depletion of energy and therefore compromised growth, and even death by a perceptive predator. Young animals know how to play from the get-go, and when they don't, we take that as a sign there's possibly something wrong.

When animals play, we can feel their deep joy. Play is contagious, and other animals feel the joy and glee as well. In her book *On Talking Terms with Dogs*, Norwegian dog trainer Turid Rugaas refers to play signals as "calming signals." Animals typically play only when they're relaxed, so the inherent joy and serenity in play often spread to anyone who is watching.

What we can see with our eyes is also being borne out by scientific research. Studies of brain chemistry in rats support the idea that play is pleasurable and fun. Renowned neurobiologist Jaak Panksepp discovered a close association between opiates and play in rats — an increase in opioid activity facilitates playfulness. Opioids may enhance the pleasures and rewards associated with playing. If this is true in rats, and we already know it's true in humans, then there's little reason to believe that the neurochemical basis of play-inspired joy in dogs, cats, horses, and bears would differ substantially.

As we've already seen, animals and humans share many of the same chemicals that play a role in the experience and expressions of emotions such as joy and pleasure. Research has also shown that when people are cooperating and being fair to one another, it feels good. Since play is dependent on cooperation and fairness, this may be another reason animals love to play.

James Rilling and his colleagues have used functional magnetic resonance imaging (fMRI) on humans to show that the brain's pleasure centers are strongly activated when people cooperate with one another. This important research shows that there's a strong neural basis for human cooperation: It feels good to cooperate, and being nice in social interactions is rewarding. Also, researchers have identified a "trust center" in human brains called the caudate nucleus. Activity in the caudate nucleus is greatest when generosity is repaid with generosity. There's every reason to believe that the brains of animals share this trust center with us. In short, research is showing that we might actually be wired to be nice to one another.

If being nice feels good, then that's a good reason for being nice. It's also a good way for a pattern of behavior to evolve and to remain in an animal's arsenal.

It Begins with a Bow: How Animals Play

In my studies of dogs and other canids, I've learned that they use specific play signals to initiate and to maintain social play. Play is a voluntary activity, and it can't occur if individuals don't agree. Dogs frequently tell one another *I want to play with you* with a bow (which I describe in chapter 2), and bowing is repeated during play sequences so as to insure that play doesn't slip into something else, like fighting or mating. After each individual agrees to play, there are ongoing, rapid, and subtle exchanges of information so that their cooperative agreement can be fine-tuned and negotiated on the run, which keeps the activity playful. It's important for players to express and to share their intentions to play.

As I've noted, when animals play, they often use actions that are also carried out in other contexts, such as predatory behavior, antipredatory behavior, and mating behavior. Because there's a chance that various behavior patterns performed during ongoing social play can be misinterpreted as real aggression or mating, individuals use a bow to tell one another such messages as *I want to play*, *This is still play no matter what I am about to do*, and *This is still play regardless of what I just did to you*.

In order to learn the dynamics of play, it's essential to pay attention to subtle details that can be lost or go unnoticed when, for instance, we are simply watching dogs in the park. Dogs and other animals keep close track of what's happening, so we need to also. My studies of play are based on careful observation and meticulous analyses of videos. My students and I watch tapes one frame at a time to see what the animals are doing and how they exchange information about their intentions and desires during play. This can be tedious work, and indeed, some of my students who were initially excited about studying dog play had second thoughts after seeing what it entailed.

After many years of study, I discovered that the "bow" isn't used randomly but with a purpose. For example, biting accompanied by rapid side-to-side shaking of the head is performed during serious aggressive and predatory encounters, and it can easily be misinterpreted if its meaning isn't modified by a bow. I was surprised to learn that bows are used not only right at the beginning of play to tell another dog *I want to play with you*, but also right before biting accompanied by rapid side-to-side head-shaking, as if to say, *I'm going to bite you hard but it's still in play*. Bows are also used right after vigorous biting, as if to say, *I'm sorry I just bit you so hard, but it was play*. Bows serve as punctuation, an exclamation point, and to call attention to what the dog wants.

In a 2016 study of pairs of adult dogs, Sarah-Elizabeth Byosiere and her colleagues discovered that play bows serve to reinitiate play after a pause rather than to mediate offensive or ambiguous actions. They also reported that 409 of 415 bows were used when the dogs could see one another. These results fit well with, and complement, what others have observed, in that bows are a sort of punctuation mark, a comma if you will, that is used strategically during ongoing play. Along these lines, in an email to me, dog enthusiast Leva Zariņa wondered if play bows could serve as "reset" signals to change the tenor of what's happening. That is, the play bow might allow the players to begin anew with little recounting of what previously happened, the change in their emotional states prompting them to treat whatever happens next as a new play bout. I like this idea and hope to see some research on this interesting suggestion.

Infant dogs and their wild relatives rapidly learn how to play fairly using play markers such as the bow, and their response to play bows seems to be innate. Pigs use play markers such as bouncy running and head twisting to communicate their intentions to play. Jessica Flack and her colleagues discovered that juvenile chimpanzees will increase the use of signals to prevent the termination of play by the mothers of their younger play partners. Researchers who study the activity always note that play is highly cooperative. I can't stress enough how important it is that play is carefully negotiated, that it is fine-tuned on the run so that the play mood is maintained. There are social rules that must be followed.

Across many different species there's little evidence that play signals are used to deceive others. Play signals are honest signals, and only very rarely are they used to hide aggressive intentions. Animals almost never say, *I want to play with you*, and then when the other animal is vulnerable, engage in a real attack. This is most likely because there are sanctions for lying. For example, I discovered that coyotes who bow and then attack are unlikely to be chosen as play partners, and they also have difficulty getting others to play. My field studies also have shown that this makes them more likely to leave their group, and this can lower their reproductive fitness.

It's simple: If a dog wants to play, they must ask first by bowing. If the other dog doesn't return the bow or jump right in, they don't want to play, and the first dog must move on.

Other Play Markers: Role-Reversing and Self-Handicapping

The bow isn't the only signal used during social play; two other important ones are role-reversing and self-handicapping. Role-reversing and self-handicapping reduce inequalities in size and dominance rank between players, and they promote the reciprocity and cooperation that's needed for play to occur.

Self-handicapping (also called "play inhibition") happens when an individual performs a behavior pattern that might compromise them outside of play. For example, a coyote might decide not to bite their play partner as hard as they can, or they might not play as vigorously as they can. Inhibiting the intensity of a bite during play helps to maintain the play mood. I once picked up a twenty-two-day-old coyote only to have him bite through my thumb with his needle-sharp teeth. His bite drew blood and it really hurt. The fur of young coyotes is very thin, and as I found out, an intense bite results in much pain, as evidenced by the high-pitched squeals (the coyotes', not mine!). An intense bite is a play-stopper. In adult wolves, a bite can generate as much as fifteen hundred pounds of pressure per square inch, so there's a good reason to inhibit its force. I once foolishly tried to show a captive adult male wolf, Lupey, where his food was by pointing toward it, and he immediately showed me that he knew where it was by clasping his mouth over my extended forearm and squeezing ever so gently. I wore Lupey's teeth marks for two weeks, but he didn't break skin; we may not have been "playing," but he inhibited his bite anyway. With domestic dogs, one of the great advantages of making sure puppies play with other puppies is that they learn bite inhibition and rarely if ever harm another individual.

Red-necked wallabies, kangaroos of a kind, engage in self-handicapping as well. Biologist Duncan Watson and his colleagues found that these playful creatures adjust their play to the age of their partner. When a partner is younger, the older animal adopts a defensive, flat-footed posture, and pawing rather than sparring occurs. Also, the older player is more tolerant of their partner's tactics and takes the initiative in prolonging interactions.

Fairness and trust are important in the dynamics of playful interactions in rats as well. Psychologist Sergio Pellis discovered that sequences of rat play consist of individuals assessing and monitoring one another and then fine-tuning and changing their own behavior to maintain the play mood. When the rules of play are violated, when fairness breaks down, so does play.

Role-reversing happens when a dominant animal performs an action during play that wouldn't normally occur during real aggression. For example, a dominant wolf might never roll over on their back during fighting, but they will do so while playing. In some instances, role-reversing and self-handicapping occur together — a dominant wolf might roll over while playing with a subordinate dog and at the same time inhibit the intensity of a bite. Self-handicapping and role-reversing, along with play invitation signals, serve to communicate an individual's intention to play, and they are important in maintaining fair play.

Fair Is Best: The Benefits of Play

Play isn't an idle waste of time. Play is essential for an individual's mental and physical well-being. Play is brain food because it provides important nourishment for brain growth; it actually helps to rewire the brain, increasing the connections between neurons in the cerebral cortex. Play also hones cognitive skills, including logical reasoning and behavioral flexibility — the ability to make appropriate choices in changing and unpredictable environments. Along these lines, two colleagues and I proposed that play provides "training for the unexpected." Based on an extensive review of available literature, my colleagues Marek Spinka, Ruth Newberry, and I proposed that play functions to increase the versatility of movements and the ability to recover from sudden shocks, such as the loss of balance and falling over, and to enhance the ability of animals to cope emotionally with unexpected stressful situations. To obtain this "training for the unexpected," we suggested that animals actively seek and create unexpected situations in play and actively put themselves into disadvantageous positions and situations.

But some of the most important benefits of play are social — play helps the individual and the group to get along together. Social play relies on, and also teaches, trust, cooperation, niceness, fairness, forgiveness, and humility. Dogs and their relatives aren't alone in the tactics they use in play. Research on nonhuman primates has shown that punishment and apology play important roles in maintaining cooperation.

Why do animals carefully use play signals? Why do they engage in self-handicapping and role-reversing? It's plausible to argue that during social play, immature individuals learn ground rules about what behavior patterns are acceptable to others — how hard they can bite, how roughly they can interact — and how to resolve conflicts in a situation that is safe, enjoyable, and nonthreatening. This is similar to the reasoning behind why human children are encouraged to play organized sports: It teaches them how to behave, how to cooperate and resolve conflicts in a setting where the stakes are not high. Through their behavior, animals show us that they place a premium on playing fairly and trusting others to do so. There are codes of social conduct that regulate actions that are and aren't permissible. What could be a better atmosphere in which to learn about the social skills underlying fairness and cooperation than during social play, where there are few penalties for transgressions? It's also possible that individuals might generalize codes of conduct learned while playing with specific individuals to other group members and to different situations, such as sharing food, defending resources, grooming, and giving care.

Play is not only fun. It's a useful behavior. And studies of play indicate that animals actively cultivate a sense of fairness and cooperation by playing. This becomes even clearer in instances where play breaks down.

A Bad Reputation: The Costs of Breaking Trust

When animals are not having fun and relaxed, they won't play. Animals who don't play often can't interact with others because they don't know how to tell their friends what they want and they can't understand what their friends want. They're not socialized. They can't function as card-carrying members of their own species because they haven't learned how to communicate with others. The consequences of the inability to play well start small but possibly grow quite large.

For instance, dogs don't tolerate noncooperative cheaters, who may be avoided or chased from play groups. When a dog's sense of fairness is violated, there are consequences. While studying dog play on a beach in San Diego, California, for her doctoral dissertation in cognitive science, Alexandra Horowitz observed a dog she called Up-ears enter into a play group and interrupt the play of two other dogs, Blackie and Roxy. Up-ears was chased out of the group, and when she returned, Blackie and Roxy stopped playing and looked off toward a distant sound. In a fooling behavior, Roxy began moving in the direction of the sound, and Up-ears ran off following their line of sight. Having gotten rid of Up-ears, Roxy and Blackie immediately began playing once again.

Of much more importance to biologists, however, is how differences in the performance of a given behavior, such as play, influence an individual's reproductive success. Do differences in play and variations in fair play affect an individual's reproductive fitness? If we want to know whether a sense of fairness or morality evolved because it's adaptive in its own right — because it improves an individual's, and thus a species', chance for survival — then we should be able to show that more "virtuous" individuals are more fit and have more offspring than less virtuous individuals (as Darwin indicated). If play and fairness are inextricably linked (as they seem to be), then is it true that individuals who play well do better reproductively than those who don't? It's almost impossible to directly link fair play with an individual's reproductive success or fitness, but it's also extremely difficult to show with great certainty that the performance of most behaviors is directly and causally coupled to reproductive success. However, my students and I have collected some intriguing data on wild coyotes that indicate a relationship between play and fitness.

Dogs, coyotes, and wolves are fast learners when it comes to fair play. There are serious sanctions when they breach the trust of their friends, and these penalties might indeed become public information if others see an individual cheating on their companions. Biologists call these penalties "costs," which means that an individual might suffer some decline in their reproductive fitness if they don't play by the "golden rules of fair play."

The Golden Rules of Fair Play

Dogs and other animals tend to follow the golden rules of fair play. They stick to mutually agreed upon codes of conduct when they zoom around or wrestle with one another. Among dogs, the moral landscape of play is *Don't bow if you don't want to play*. Here are the four golden rules of play:

1. ASK FIRST AND COMMUNICATE CLEARLY

Many nonhumans announce that they want to play and not fight or mate. Canids punctuate play sequences using a bow to solicit play, crouching on their forelimbs while standing on their hind legs. Bows are used almost exclusively during play and are highly stereotyped — that is, they always look the same — so the message *Come play with me* or *I still want to play* is clear. Play bows are honest signals, a sign of trust.

Even when an individual follows a play bow with seemingly aggressive actions, such as baring teeth, growling, or biting, their companions demonstrate submission or avoidance only around 15 percent of the time, which suggests they trust the bow's message that whatever follows is meant in fun. Trust in one another's honest communication is vital for fair play and a smoothly functioning social group.

2. MIND YOUR MANNERS

Many animals consider their play partners' strength, size, and physical abilities and engage in self-handicapping and role-reversing to create and maintain equal footing. For instance, a coyote might not bite their play partner as hard as they can, handicapping themselves to keep things fair. And a dominant pack member might perform a role-reversal, rolling over on their back (a sign of submission that they would never offer during real aggression) to let their lower-status partner take a turn at "winning."

3. ADMIT WHEN YOU ARE WRONG

Even when everyone wants to keep things fair, play can sometimes get out of hand. When an animal misbehaves or accidentally hurts their play partner, they typically apologize, just like a human would. After an intense bite, a bow sends the message: *Sorry I bit you so hard — this is still play regardless of what I just did. Don't leave; I'll play fair.* For play to continue, the other individual must forgive the wrongdoing. And forgiveness is almost always offered; understanding and tolerance are abundant during play as well as in daily pack life.

4. BE HONEST

An apology, like an invitation to play, must be sincere. Individuals who continue to play unfairly or send dishonest signals often quickly find themselves ostracized. There are far greater consequences than simply reduced playtime.

Blackballed from the Group

I was the first to offer a possible "big picture" evolutionary reason for the importance of admitting to having messed up and violated the agreed-upon golden rules of play and social conduct. For example, my long-term field research on coyotes living in Grand Teton National Park outside of Moose, Wyoming, showed that juvenile coyotes who do not play fair often ended up leaving their pack either because their play invitations were ignored or they themselves were avoided, rather than being forced out by siblings or other group members. These coyotes left the pack because they were unable to establish strong emotional bonds with other group members. Then, because life outside the group can be tougher, they were up to four times more likely to die than those individuals who remained with others. There are substantial risks associated with dispersal by young coyotes, and violating social norms established during play is not good for perpetuating one's genes.

Our fieldwork on coyotes revealed one direct cost paid by animals who fail to engage in fair play or who don't play much at all. Were these outcomes totally related to the way these coyotes played? We're not sure, but information that we collected on captive coyotes suggested that the lack of play was a major factor in individuals spending more time alone, away from their littermates and other group members.

While more research can and should be done, it seems quite likely that breakdowns in social play negatively affect individuals, and by extension their social groups. In social species at least, natural selection seems to weed out cheaters, those who don't play by the accepted and negotiated rules. Conversely, animals, including humans, survive and thrive better when they play fair and learn the group's moral codes for behavior. In other words, it seems like morality evolved because it's adaptive.

Play Is Intelligent, Egalitarian, and Just

All in all, fair play can be understood as an evolved adaptation that allows individuals to form and maintain social bonds. The parallels between human and animal play, and the shared capacity to understand and behave according to rules of right and wrong conduct, are striking. Canids, like humans, form intricate networks of social relationships and live by rules of conduct that maintain a stable society, which is necessary to ensure the survival of each individual. Basic rules of fairness guide social play, and similar rules are the foundation for fairness among adults. It may have been just this sense of right and wrong that allowed human societies to flourish and spread across the world.

Successful play also takes intelligent cognitive and emotional awareness. Dogs keep track of what is happening when they play — how they're feeling and how their playmates are feeling — and they can be quite creative on the go, in part because it's in the best interests of all players that fairness remains the name of the game. Dogs must be able to read the intentions of other dogs, and they trust dogs when they indicate they want to play, not fight. That means dogs need to conceive of what others are thinking and feeling. They need to pay close attention to what other dogs have done and are doing, and they use this information to predict what other dogs are likely to do next. That strongly suggests that dogs probably do have a theory of mind — that is, the ability to know another individual's thoughts and feelings. In fact, research on dog play is one of the main ways we've discerned this.

A few people have asked me if dogs *always* play fair, and they mention examples in which play among dogs escalated into what *seemed* to be an aggressive encounter, or it *seemed* like that was about to happen. I explain that this is very rare. People often confuse rough-and-tumble play fighting with real aggression. Among the thousands of playful interactions I've observed in dogs in different locations, I remember watching three dogs playing during which they jumped on one another's backs while biting and head-shaking very vigorously. The guy who was with them told me that they played like this all of the time and never once had it escalated into an assertion of dominance. To the untrained eye, it looked as if they really were beating the hell out of one another. I've seen interactions like this among young and old dogs in dog parks and among wild coyotes, wolves, and red foxes. When left on their own, dogs and many other animals have a way of sorting things out as they engage in rough-and-tumble play.

The bottom line is that, although play is fun, it's also serious business. When animals play, they are constantly working to understand and follow the rules and to communicate their intentions to play fairly. They fine-tune their behavior on the run, carefully monitoring the behavior of their play partners and paying close attention to infractions of the agreed-upon rules.

It may turn out that play is a unique category of behavior in that inequalities are tolerated more so than in other social contexts. Play can't occur if the individuals choose not to engage in the activity, and the equality or fairness needed for play to continue makes it different from other forms of cooperative behavior (such as hunting and caregiving). Play is perhaps uniquely egalitarian, and this sort of egalitarianism in animals is thought to be a precondition for the evolution of social morality in humans. If we define justice or morality as a set of social rules and expectations that neutralize differences among individuals in an effort to maintain group harmony, then that's exactly what we find in animals when they play.

Survival of the Friendliest

> *I believe that at the most fundamental level our nature is compassionate, and that cooperation, not conflict, lies at the heart of the basic principles that govern our human existence.... By living a way of life that expresses our basic goodness, we fulfill our humanity and give our actions dignity, worth, and meaning.*
>
> — HIS HOLINESS THE DALAI LAMA,
> "UNDERSTANDING OUR FUNDAMENTAL NATURE"

> *We believe that most donkeys, if given the chance, would fashion a world without violence. Like St. Francis of Assisi, they would remake the natural world into a proverbial garden of Eden, where the lion and the lamb lay side by side.... We wonder how much the innocence, vulnerability, and gentleness of [those two] donkeys may have affected Tolstoy, the peace-loving giant of world literature.*
>
> — MICHAEL TOBIAS AND JANE MORRISON, *DONKEY*

Some might call the Dalai Lama an optimist, but increasingly, scientific data support His Holiness. Indeed, the time has come to revise our notions about what "survival of the fittest" really means.

So far, most of the research on cooperation has focused on humans. As it turns out, humans aren't as selfish and self-centered as we're sometimes made out to be. Ernst Fehr and his colleagues have discovered that when treated fairly, many people will voluntarily cooperate with one another and also punish those who don't cooperate. They call this "strong reciprocity," and they show that it can lead to "almost universal cooperation in circumstances in which purely self-interested behavior would cause a complete breakdown of cooperation." They note that people are willing to punish individuals who behave unfairly to a third person.

There's also evidence that humans have a natural tendency to be altruistic. Felix Warneken and Michael Tomasello discovered that infants as young as eighteen months of age will help people in need, such as when they're searching for a lost object. Young chimpanzees also will do this. What's very interesting about this study is that young children, while still in diapers and not very skilled in using language, will only help retrieve a lost object when they believe that a person needs the object to complete a task. For example, they would only retrieve a clothespin if it seemed to have been dropped unintentionally by the researcher, but not if it was clearly thrown on the ground deliberately. Children can understand through body language when something is needed or not, and they help only if there is a need.

Research with animals shows similar findings. Obviously, as we've seen throughout, when it comes to play, cooperation and trust are required, and almost every piece of communication is aimed at maintaining that state of affairs. But cooperation seems to predominate in a wide range of social situations. Primatologists Robert Sussman and Paul Garber report that for diurnal prosimians, such as lemurs, New World monkeys, and great apes, the vast majority of social interactions are affiliative rather than agonistic or divisive. Grooming and bouts of play predominate in the affiliative category. For the prosimians, an average of 93.2 percent of social interactions are affiliative, and the numbers for the New World monkeys and Old World monkeys are 86.1 and 84.8 percent, respectively.

Clearly, if these numbers can be repeated among other species, competition does not drive animal behavior; cooperation and friendliness aren't simply sideshows to aggression and fighting. Indeed, it almost seems trite to write about cooperation in animals because everyone knows that animals cooperate with one another, and it's obvious why: If the group works together, then each individual's chance for survival improves.

In their landmark 2020 book *Survival of the Friendliest*, Brian Hare and Vanessa Woods argue that the secret to our success as a species is our unique friendliness. They maintain that cooperation, trust, and friendliness have played critical roles in the evolution of social behavior and group living and are widespread among diverse animals. All three attributes go hand in hand and are the basic ingredients of forming and maintaining cooperative and cohesive groups, packs, flocks, or herds of animals. Conversely, lying, cheating, and deceiving undermine trust among individuals and weaken, if not break, social bonds. Nor does trust require language. In other words, they say, developing the attributes of friendliness is what makes animals the most fit.

However, much of evolutionary theory is still based on competition among individuals rather than on cooperation, and for some people this has meant that cooperation must be a by-product of competition that is not directly selected for in evolution. In this view, animals only cooperate because, if they were competitive all of the time, it would be difficult to form and maintain stable social groups. For example, it would be impossible to have a stable wolf pack in which all of the males were top wolf or alpha individuals who always competed with one another. Cooperative interactions have to back competitive encounters so that the group will be cohesive and stable over time.

This "survival of the fittest" mentality, which still pervades so much thinking and theorizing, is increasingly not supported by research as being the prime mover in evolution. For a long time, cooperation has been ignored because of this ideological bias, but a deluge of research papers and essays on cooperation indicate that the tide is changing. In fact, the more we look for cooperation, the more we discover its presence. Animals certainly still compete, but cooperation is central in the evolution of social behavior, and this alone makes it key for survival. When animals cooperate, they're doing what comes naturally, and cooperation relies on established, well-maintained social standards of behavior — that is, moral codes. This is what should become the starting point for evolutionary theory and the basis for our discussions about the lives of animals.

Some ecologists take this even further. They wonder if, when studying "ecological interactions" — that is, encounters among different species of animals and interactions between animals and trees and plants — it makes more sense to concentrate on positive ecological interactions rather than on competition and predation. These researchers have been called "renegade ecologists" by more mainstream scientists, and they argue that there is more to ecology and the evolution of communities than just competition and predation. They maintain that a process called "facilitation" readily works alongside competition and provides a balance to its mechanisms that is important in the evolution of community structure. However, it's one thing to say that cooperation within a group of individuals, for example, a wolf pack or a troop of baboons, is beneficial — that all wolves benefit more by helping one another than by competing with one another — but is cooperation an essential component of an entire forest or ecosystem? It's an intriguing notion, one on which I'm sure His Holiness the Dalai Lama would like to see more research.

Is There a "Universal" Morality?

Justice presumes a personal concern for others. It is first of all a sense, not a rational or social construction, and I want to argue that this sense is, in an important sense, natural.

— Robert Solomon, *A Passion for Justice*

It's self-serving anthropocentric speciesism to claim that we're the only moral beings in the animal kingdom. Social morality, as exhibited during play, is an adaptation that is shared by many animals. Behaving fairly evolved because it helped young animals acquire social (and other) skills needed as they matured into adults. Being fair is also important for maintaining well-oiled and efficient cooperative groups. There may even be cooperation between different species during group hunts. If we find consistency among different species in terms of how they cooperate and negotiate agreements to be fair, we might discover a universal morality. Such morality might also be important in acquiring, defending, and sharing food, in social grooming, and in the communal care of youngsters.

In *The Origins of Virtue*, biologist Matt Ridley points out that humans seem to be inordinately upset about unfairness. However, we don't know much about how other animals react to unfairness. Nonetheless, we are discovering some good leads to build on from observations of animals at play — and we may soon be able to include a sense of injustice as a shared trait. What about forgiveness? This is another moral sense that is often attributed solely to humans, but the renowned evolutionary biologist David Sloan Wilson has shown that forgiveness is a complex biological adaptation. In his book *Darwin's Cathedral: Evolution, Religion, and the Nature of Society*, Wilson says, "Forgiveness has a biological foundation that extends throughout the animal kingdom." And further, "Forgiveness has many faces — *and needs to* — in order to function adaptively in so many different contexts" [Wilson's emphasis]. While Wilson concentrates mainly on human societies, his views can easily be extended — and responsibly so — to nonhuman animals. Indeed, Wilson points out that adaptive traits such as forgiveness might not require as much brain power as once thought. This isn't to say that animals aren't smart, but rather that forgiveness might be a trait that is basic to many animals, even if they don't have especially big and overworked brains.

It's clear that morality and virtue didn't suddenly appear in the evolutionary epic beginning with humans. The origins of virtue, egalitarianism, and morality are more ancient than our own species. While fair play in animals may be a rudimentary form of social morality, it still could be a forerunner of more complex and more sophisticated human moral systems. But perhaps most important, if we try to learn more about forgiveness, fairness, trust, and cooperation in animals, maybe we'll also learn to live more compassionately and cooperatively with one another.

{ 5 }

Addressing Uncertainty and Anthropomorphism in the Study of Animal Minds

Sometimes I read about someone saying with great authority that animals have no intentions and no feelings, and I wonder, "Doesn't this guy have a dog?"

— Frans de Waal, "A Conversation with Frans de Waal"

I also ask the same question as Frans de Waal — I'm incredulous when someone denies what's self-evident in every household with a dog. Ever since early childhood, I've wondered, *What is it like to be a fox?* or *What does it feel like to be a fox?* Through high school, college, and beyond, this interest remained, and when I discovered the field of cognitive ethology, I knew that this is what I wanted to study. My parents told me that I always "minded animals." I now understand this in two ways: I always attributed minds to animals, but I also minded them — cared for them, respected them, and loved them. This has been a central part of who I am since childhood. It's innate — my evolutionarily "old brain" pulls me back to animals and to nature. Even today I ask the same question: What is it like to be a dog, or a wolf, or a coyote? The difference is that today I have the training and experience to actually find out, and it is a great satisfaction to feel I've made a lot of progress in answering this question. Of course, even after five decades of studying animal behavior and cognitive ethology, there is still more to learn, and I'm as curious as ever.

A lot of my early animal education came from the warm and compassionate home in which I was raised. There also was a healthy dose of positive thinking, and this combination was critical for setting me on the path I've been traveling, with a few detours here and there, for my entire life. As a child, I always felt for other animals — I honestly could feel and internalize their joy, sorrow, and pain. It was natural for me to empathize with them, and today, I am indeed sorry for any harm that I have caused to animals in the course of my research and daily activities. I know I'm not perfect, but I try as hard as I can to do no harm — to be proactive, to anticipate the various stresses that different types of research might cause for animals. I think it's important to voice your beliefs, and so I take part in peaceful and compassionate protests on behalf of animals, from wild animals like black-tailed prairie dogs and Canadian lynx to domesticated animals in pet stores, puppy mills, kitten factories, labs, and on food farms. On more than one occasion I've ruffled a few academic feathers by making my views known in scientific circles where they're not always accepted. I also believe that scientists are responsible for sharing their findings with nonscientists. Scientists need to be less self-absorbed and more community and globally minded.

I tell you all this because it's important to stress that science isn't value-free. Each scientist brings to their work a certain set of values; these influence how we conduct research and explain and interpret data. The goal of science is to reach "objective" conclusions about the world — answers that are free of personal bias — but scientists themselves are not unfeeling automatons. They're individuals, real people with particular points of view. Science has always struggled with this. At what point does subjective knowing become objective "truth"? How much research, and what kind, needs to be done to prove something? How much does a scientist's beliefs influence how they interpret "objective" data without being aware of this bias? Do a researcher's intuitions and feelings, the personal self, ever have a place in science?

These are the issues that I want to address in this chapter. They are the "hard questions" that all scientists must ask of themselves and their work, but they are particularly relevant to cognitive ethology, which relies on anecdotes, analogy, and anthropomorphism to reach its conclusions. These have traditionally been "dirty words" in science, since they smack of the subjective and the personal; invariably, when scientists criticize ethology and its findings, one or all of them are mentioned. But are people who resist these three A words themselves reacting out of personal or professional bias? Could there be a place in science for subjectivity that doesn't compromise "objective truth"? Most pointedly of all, could the discomfort some scientists have with acknowledging animal emotions reflect, not the quality of the evidence, but fear of being seen as "unscientific"? Personally, I think that stepping out of one's comfort zone is a great educational motivator. A healthy dose of common sense also can go a long way.

I love being a scientist and conducting scientific research, but I remain open to other ways of knowing, to innovative thinking. I don't think it's a matter of science *or* subjectivity but rather science *and* subjectivity. Personal truths are valid; if they are acknowledged and accounted for, they don't need to compromise objectivity. As I explore further below (regarding anthropomorphism), it may very well be that emotions are not strictly personal, either, but they may have evolved as social adaptations, tools for understanding what others are feeling. In any case, we need to be able to live with uncertainty. Many scientists like the autonomy and authority society grants them, and they are loath to jeopardize their status by seeming too personal or unsure, but we must all give up control. Science and scientists must be dynamic, open, and compassionate.

Much of science is still written in the third person rather than the first person. The use of the third person — "the researcher did this" or "the subjects were observed by the researcher" — makes it easy for researchers to distance themselves from the animals they study, and even from the process of conducting their research. Third-person language can have a touch of arrogance — of the "Back off, I'm the scientist" variety — and it allows researchers to pretend that their personal values and perspectives don't influence their results. It reinforces the erroneous notion that science is value-free. Most important, third-person language denies not only the subjectivity of the scientist but the subjective lives of the animals being studied. I'm pleased to say that things have been changing. Today, some scientific journals allow researchers to write in the first person, and use the pronouns "I" and "we," and also to name, rather than number, the animals they study.

Animals are subjective beings, but scientists frequently treat them like objects because that is what they are trained to do. This does a disservice to animals, to scientists, and to our knowledge of one another. If we choose to study other animals, as well as speak for them, then we should represent them with open hearts and clear minds. Our research should be both diligent and compassionate. The study of animal behavior, animal emotions, and research about the nature of human-animal interactions would surely benefit from, if not require, a more honest and open first-person approach.

Bill and Reno: Certain at Home, Uncertain at Work

When it comes to animal sentience, consciousness, and emotions, some scientists have a hard time reconciling how they act with their own pets at home with what they believe and do at work. When it comes to their pets, they treat them with the love and affection of a family member, marvel at their games and knowledge, and feel the satisfaction of being loved in return. Then they go to work, and suddenly they just can't say for sure if animals feel or know anything. This is so backward I'm tempted to make the subtitle for this story "Correct at Home, Wrong at Work."

I've experienced the double life of researchers many times, as they shift from being at work to being out with their dogs and having a beer, but the following story of Bill, a professor I knew, and his dog Reno is a classic. Years ago, before a lecture I was going to give at a major American university, Bill came up to say hello. I asked him how Reno was doing, and for more than five minutes, Bill told Reno stories — how Reno loves to play with his friends, how he anxiously misses Bill when he's gone, and how a few days before Reno became jealous of the attention Bill was giving his daughter. Reno sounded like a pretty bright and emotional dog. Bill described him being happy and sad, treasured Reno's love for his human family, and illustrated how vividly and clearly Reno communicated all this to Bill.

I was thrilled to hear these stories, as Bill had always been dubious about animal emotions. The furthest he would go was to say that animals acted "as if" they experienced emotions, but it was too early to make any grandiose pronouncements about what their feelings were or if they had them at all.

My excitement was short-lived. During the question-and-answer session following my talk, Bill accused me — in a sort of lighthearted academic way — of being far too anthropomorphic and too sure of myself. I was taken aback, but instead of arguing the point, I simply asked Bill to recall for the audience the conversation that we'd had before my lecture about Reno's emotional life. Bill turned slightly red, and then said, "Well, you know what I meant when we were talking before. I was just letting my hair down and telling stories about Reno. I really don't know that he enjoys playing with his friends or that he really is depressed when I leave him alone. And I feel rather certain that he wasn't really jealous of the attention I was giving to my daughter. He just behaved as if he were."

To be frank, I still don't really know what Bill meant. He was quite comfortable telling me about Reno's feelings in one context but not in another.

Did he mean to say he lived in a fantasy world at home (in which he preferred acting "as if" his dog possessed emotions), but returned to "reality" at work? Or did he mean that he was reluctant to behave in his job in ways, or to espouse beliefs, that were consistent with his personal experience? For whatever reason, Bill couldn't reconcile his personal and scientific selves, and he didn't even seem to recognize that there was a disconnect — or dissonance, as psychologists call it — between what he told me an hour earlier and what he said he believed during the lecture.

Similar conversations among scientists take place at cocktail parties when it's permissible to let your guard down — when scientists don't feel they need to sanitize animal emotions with all sorts of "Well, you know what I mean" disclaimers. In these settings, scientists will speak freely of their pets' intelligence and feelings, but those statements vanish once they put on the lab coat Monday morning. This distancing mechanism seems essential to allow them to conduct their work, but does it make for good science or healthy scientists? If a scientist is conducting experiments that harm an animal they would love in another context, like a dog, wouldn't it be hard for the person to admit the animal was feeling and realizing everything the person was doing to them? In this case, wouldn't "doubt" about animal emotions simply be a way for the scientist not to feel so bad? Or is "doubt" sometimes a fig leaf for scientists who don't want to commit the sin of bringing personal emotions to their work — and suffering the possible criticism or judgment of colleagues? I didn't have the heart to ask Bill if he would do to Reno what he does to other animals in his lab. I expect he wouldn't have.

In cases like Bill's, my suspicion is that doubt about animal emotions is espoused not because it serves science but because it protects the emotional needs of the scientist.

Asleep on the Job: How Do You Know What You Say You Know?

Bill was pretty nice about our disagreements. However, some people get more upset, and their skepticism is profound. Right after I gave a lecture on animal emotions in Bochum, Germany, a woman in the audience — who looked as bored as bored could be — stood up and accused me and other fieldworkers of always "being sleep deprived," and that's why we attribute active minds and deep feelings to animals. I tried to disarm her because I thought she was partly joking (and also because I really was sleep deprived, from jet lag!), but she became rather hostile. She was quite serious, and I wasn't sure what to say. So I answered the charge seriously: I told her that cognitive ethologists were scientists and serious researchers who didn't sleep on the job. Even if arduous fieldwork did make us tired, this wasn't the reason we wrote about animal passions. We wrote about them because we observed them in the field, and we have learned they've evolved as important adaptations. Fieldwork isn't perfect; it has benefits as well as shortcomings. Lab-bound researchers (who assign numbers to animals they study in cages) often criticize fieldworkers for being too lax and for not having control over the animals they observe. But I counter this by noting that we can gain the most complete understanding of the rich, complex lives that animals lead by observing them in their natural environments. In any case, fatigue and a lack of control can't conjure emotions out of thin air.

Similarly, I recall an event at a symposium that was held at the Smithsonian Institution in 2000. Cynthia Moss talked about her long-term research on the social behavior of wild elephants in Kenya, and she showed a wonderful video of these highly intelligent and emotional beasts. During the question-and-answer period, a former program leader from the National Science Foundation asked Cynthia, "How do you know these animals are feeling the emotions you claim they are?" And Cynthia aptly replied, "How do you know they're not?" Of course, neither question could be answered with absolute certainty, but Cynthia had been studying wild elephants for decades rather than caged, bored, and stressed individuals. Her own and a lot of other scientific research overwhelmingly supports Cynthia's view. For decades, despite our best efforts, there has remained an element of doubt, even if it's rapidly evaporating.

When it comes to the interior life of any other being — including humans, but particularly beings of another species — there may always be a point beyond which we cannot see or measure or know. The presence of some doubt doesn't call into question the whole field of cognitive ethology, however, nor does it mean that hopeful researchers are engaging in wishful thinking or spinning exhaustion-fueled fantasies. Cognitive ethologists are sincere scientists; they continue to bring rigor to a difficult area of study and to put forth solid explanations of the evidence they find.

The study of animal emotions is strongly evidence based, and these days, people who do hard science are generating more and more of the data. One reason the study of animal emotions is so exciting is that the down-home observations of animals doing what they do are being supported by laboratory-bound researchers who have chosen to study the neural basis for emotions and even why humans anthropomorphize.

Speaking Out: Being Passionate about Animal Passions

Most scientists have traditionally been hesitant to speak freely about animal emotions, but times are changing. When my book *The Smile of a Dolphin: Remarkable Accounts of Animal Emotions* was published in 2000, it was a coming-out party for more than fifty distinguished scientists who told their own stories about the emotional lives of the animals they knew best — the animals with whom they had bonded during years of study. They told stories about love in dolphins, fish, and dwarf mongooses; about anger and aggression in octopuses, ravens, and cats; of joy and grief in spotted hyenas, elephants, and rats; and of embarrassment, resentment, and jealousy in primates. These stories demonstrate that many different animals have rich emotional lives, and they also make clear that the scientists themselves have feelings for the animals they study.

While this may not sound revolutionary today, or to nonresearchers who care about animals, back then, for scientists to display their feelings and to write freely about animal emotions, to speak about individuals they had named rather than numbered — that was a big step forward. Events like this helped to legitimize the formal study of animal passions as well as legitimize the passions of ethologists. Astronomers can speak glowingly and poetically about the night sky, sharing their love for the stars, without fear that their emotions will lead anyone to doubt the reliability of their data. People understand that passion is essential; it keeps an individual going when problems arise and the hours grow long. Passion fuels the curiosity essential for scientific inquiry. And people understand that astronomers can be both passionate and meticulously accurate at the same time. Even today, ethologists do not always enjoy that same presumption.

Because of limited space, I wound up turning down the requests of a number of distinguished scientists who wanted to tell their stories about animal emotions in *The Smile of a Dolphin*. But scientists themselves continue to break down barriers, and many books, journals, publications, and venues now provide opportunities for scientists to discuss feelings, their own and those of the animals they study. As I mention, researchers no longer have to sanitize their prose or play verbal gymnastics about their claims. Scientists are free to tell their stories, and the implications of their detailed observations for how we need to treat animals are obvious to all.

Anecdotes: The Importance of Stories

When reviewing someone's research, scientists always consider how the data were collected, whether the data seem reliable, and how the data are finally explained, interpreted, and disseminated. Anecdotes, or stories, are a type of data, and they always find their way into people's descriptions of animals. However, some scientists dislike or ignore anecdotes because they're "merely stories." They're not "hard data" because they're not reproducible, and they're potentially too tainted with personal involvement and biases.

However, much of our theorizing about the evolution of behavior also rests on stories. Few scientists find this objectionable, perhaps because these stories revolve around the widely accepted central unifying theory of natural selection. In fact, systematic analyses of anecdotes can lead to data that are reproducible by organizing experiments that mimic the anecdotal situations.

I like to say that the plural of *anecdote* is *data*. Anecdotes are central to the study of animal behavior and animal emotions, as they are to much of science, and rightfully so. Emotions don't occur in a vacuum. They occur in context; there are events that cause them, and consequences that follow, and to properly describe emotions requires telling a story. How do we know a baboon was angry? Well, he was eating, someone took his food, he screeched at and chased the other baboon, and then took his food back. How do we know a young red fox missed and loved his mother? He squealed, searched for her, and when the two were reunited, he snuggled her closely, closed his eyes, and quietly fell asleep. We could restrict our description to just the animals' actions and manner, but it's the larger context that helps us identify their intentions and the emotions behind the actions. The theft and the mother's absence help clarify that the screeching/chasing and the squealing/snuggling didn't indicate something else, like joyful play or amorous mating.

As we accumulate more and more stories, we develop a solid behavior database that can be used to stimulate further empirical research, and yes, additional stories. It is important to note the frequency with which similar stories surface among different species, and these also help us identify particular emotions. Different species may express emotions using different behaviors, but the context of the story helps make clear that the emotions are similar. Further, the more stories we gather showing the same thing, the less likely it will be that personal bias has influenced the collection of data or our conclusions. In the end, anecdotes are only data that are perhaps gathered more slowly, but that doesn't make them any less useful or reliable.

The Inevitable "Sin": Anthropomorphism

It is possible, therefore, that your simple man, who lives close to nature and speaks in enduring human terms, is nearer to the truth of animal life than is your psychologist, who lives in a library and today speaks a language that is tomorrow forgotten.

— WILLIAM J. LONG, BRIER-PATCH PHILOSOPHY BY "PETER RABBIT"

In the late 1990s two remarkable novels were published: White as the Waves, a retelling of Moby Dick from the perspective of the whale...and The White Bone, about the destruction of elephant society as seen by elephants....Both novels use what is known of the biology and social lives of these species to build pictures of elaborate societies, cultures, and cognitive abilities. Their females are concerned with the survival of calves: their males inhabit a rich social and ecological fabric of which mating is only a small part. A reductionist might class these portraits with Winnie-the-Pooh as fantasies on the lives of animals. But for me they ring true, and may well come closer to the natures of these animals than the coarse numerical abstractions that come from scientific observations.

— HAL WHITEHEAD, SPERM WHALES

Along with anecdotes, anthropomorphism has frequently been used to bash the field of cognitive ethology. There are many different ways of describing what animals do. How one chooses to summarize what animals see, hear, or smell depends on the questions in which one is interested. There isn't only one correct way to describe or to explain animal behavior and emotions. In the study of animal behavior, anthropomorphism is the attribution of human characteristics to nonhuman animals. Using words that describe such human characteristics as thinking, joy, grief, embarrassment, and jealousy is to be anthropomorphic. It's a common practice that irks many of my colleagues, although most of them freely engage in it.

In fact, the issue of anthropomorphism can often be a conversation stopper. People tend to be adamant in their beliefs about it — either that it is always and inherently wrong or that it is always and inherently right — and so I don't talk about it much anymore or try to change people's minds. This debate can often deflect attention from the more important discussion about how animals are being used and abused. However, anthropomorphism is more than it seems, and it is worth understanding properly.

As the quotes above make clear, the suspicion that anthropomorphic explanations may in fact accurately describe animals has been around a long time. Perhaps William Long, writing more than a hundred years ago (*Brier-Patch Philosophy* was published in 1906), could be dismissed as being "out of touch" with current research, but Hal Whitehead is recognized as one of the world's leading whale researchers, and it's clear that he sees the value of anecdotes and anthropomorphic explanations.

As humans who study other animals, we can only describe and explain their behavior using words with which we're familiar from a human-centered point of view. So when I try to figure out what's happening in a dog's head, I have to be anthropomorphic, but I try to do it from a dog-centered point of view. Just because I say a dog is happy or jealous, this doesn't mean he's happy or jealous as humans are, or for that matter the same as other dogs. Being anthropomorphic is a linguistic tool to make the thoughts and feelings of other animals accessible to humans. And while we surely make errors from time to time, we're pretty good about making accurate predictions in the mental realm.

Konrad Lorenz, by studying the mechanisms that encourage human adults to nurture their young, also seems to have found part of the reason we're prompted to anthropomorphize. Lorenz noted that human beings are attracted to certain characteristics of juvenile animals, including a "relatively large head, predominance of the brain capsule, large and low-lying eyes, bulging cheek region, short and thick extremities, a springy elastic consistency, and clumsy movements." In other words, juvenile animals look like human babies, and this "cute response" may be further encouraged by traits that increase the seeming "cuddliness" of an animal: fur, fuzziness, and softness of skin or hair. Anthropomorphism is clearly at work when we're drawn to young animals, and I don't know of anyone who hasn't called a puppy "cute."

Addressing Uncertainty and Anthropomorphism

In scientific circles, however, anthropomorphism is frequently a big no-no, and don't try using it with a sense of humor. Just ask Professor Robert Sapolsky, who tells the story of Nick, the pissy baboon bully, in chapter 3. "Do I get grief for the fact that in communicating, say, about the baboons I'm doing so much anthropomorphizing?" Sapolsky asks. "One hopes that the parts that are blatantly ridiculous will be perceived as such. I've nonetheless been stunned by some of my more humorless colleagues — to see that they were not capable of recognizing that. The broader answer, though, is I'm not anthropomorphizing. Part of the challenge in understanding the behavior of a species is that they look like us for a reason. That's not projecting human values. That's primatizing the generalities that we share with them." In other words, we all recognize and agree that nonhumans and humans share many traits, including emotions. Thus, we're not inserting something human into animals, but we're identifying commonalities and then using human language to describe what we observe.

When we anthropomorphize, we're doing what comes naturally, and we shouldn't be punished for it. It's part of who we are. Early in her career Jane Goodall was criticized for not using scientific methods, for naming chimpanzees rather than assigning each a number, for "giving" them personalities, and for maintaining they had minds and emotions. We've come a long way since the 1960s in many areas, but unfounded fears over anthropomorphic language linger on. It's time to put them to rest, for the betterment of animals and for the betterment of science.

Among early humans, anthropomorphizing may have allowed hunters to better predict the behavior of the animals they hunted, and it's very useful for learning more about beastly passions today. Stephen Jay Gould agreed: "Yes, we are human and cannot avoid the language and knowledge of our own emotional experience when we describe a strikingly similar reaction observed in another species." Anthropomorphism endures because it is a necessity, but it also must be done carefully, consciously, empathetically, and biocentrically. We must make every attempt to maintain the animal's point of view. We must repeatedly ask, "What is that individual's experience?" Claims that anthropomorphism has no place in science or that anthropomorphic predictions and explanations are less accurate than more mechanistic or reductionistic explanations are not supported by any data. Careful anthropomorphism is alive and well, as it should be, since it helps science along.

Anthropomorphic Double-Talk:
Animals Can Be Happy, but Not Sad

I have learned that anthropomorphism *is a deeply suspect word,
used to defend cruelty to creatures unable to speak
and defend themselves against human exploitation.*

— SIR BRIAN MAY,
FOUNDING MEMBER OF QUEEN AND THE SAVE ME TRUST

I agree with Brian May. When it comes to anthropomorphic language, I've noticed a curious phenomenon over the years. If a scientist says that an animal is happy, no one questions it, but if a scientist says that an animal is unhappy, then charges of anthropomorphism are immediately leveled. Like the dissonant personal beliefs of scientists in the story involving my friend Bill, this "anthropomorphic double-talk" seems mostly aimed at allowing humans to feel better about themselves. It's no less anthropomorphic to claim animals are happy than to claim they're not.

A good example is the story of Ruby, a forty-three-year-old African elephant who once lived at the Los Angeles Zoo. In fall 2004, Ruby was shipped back to the Los Angeles Zoo from the Knoxville Zoo in Tennessee because people who saw Ruby in Knoxville felt that she was lonely and sad. A videotape recorded by Gretchen Wyler, of the Humane Society of the United States, showed Ruby standing alone and swaying. Wyler said Ruby behaved like "a desperate elephant." Sad and lonely animals often rock back and forth repeatedly. This stereotyped behavior is not normal and is characteristic of bored and distressed animals.

Stereotypies are the physical manifestations of a pathology caused by confinement. When seen in captive animals, it's called *zoochosis*. The term was first used by Bill Travers, who cofounded the Born Free Foundation; he rightly identified the abnormal repetitive, obsessive behaviors of zoo animals as a form of psychosis. Captivity literally drives animals mad. Animals in zoos can be seen pacing back and forth, tracing and retracing a particular route through their enclosure, plucking out all their feathers or pulling out all their hair, scratching or rubbing or licking themselves to the point of serious self-injury, biting the bars of their cages for hours on end, and engaging in what zoo managers euphemistically call "regurgitation and reingestion" (eating their own vomit).

Wyler and others who claimed that Ruby was unhappy were accused of

Addressing Uncertainty and Anthropomorphism

being anthropomorphic by people who thought that Ruby was doing just fine, both in Knoxville and in Los Angeles. The former director of conservation and science for the Association of Zoos and Aquariums (AZA), Michael Hutchins, claimed that it was bad science to attribute humanlike feelings to animals, saying: "Animals can't talk to us so they can't tell us how they feel." He was critical of people who claimed that Ruby wasn't doing well in captivity and was unhappy because she lived alone and had been shipped from one place to another during the previous few years, leaving her friends behind.

The mayor of Los Angeles also weighed in and claimed, "She's in good spirits, and we're glad to have her back." And John Capitanio, the associate director for research at the California Primate Research Center at the University of California–Davis, made the following claim: "Do animals have emotions? Most people are willing to say they do. Do we know much more than that? Not really." Hutchins went on to discount the view that Ruby was unhappy, saying: "An animal might look agitated, but it might not be. It might be playing. It might look like it's playing, but be quite aggressive."

Hutchins is right — it's possible to mistakenly classify an animal's behavior, but it's wrong to imply we can never figure it out. Careful and detailed behavioral studies have shown time and again that we can indeed differentiate and understand animal behavior and how it differs in various social contexts.

Does it matter whether Ruby was happy or sad? It does. If she were shown to be unhappy, the zoo would be obligated to care for her better. Zoo officials and the LA mayor felt very comfortable saying she was "doing well," and Hutchins and Capitanio felt it was "good science" to rebut any claims to the contrary. But seeing positive emotions in Ruby is as anthropomorphic as seeing negative emotions. In fact, anthropomorphism was not really the issue; the charge was merely a smoke screen to discredit the other side. The issue was animal welfare, and the only thing to decide — since we can't know with absolute certainty what Ruby was thinking — was whose interpretation seemed most likely, given what we do know of her history and of elephant behavior.

Hutchins and Capitanio did not specifically address elephant behavior; they addressed only their ideological "foes." But those who work with elephants know that one ignores an elephant's "mood" at one's own peril. British philosopher Mary Midgley said it well:

> Obviously the mahouts [elephant keepers] may have many beliefs about the elephants which are false because they are "anthropomorphic" — that is, they misinterpret some outlying aspects of

elephant behaviour by relying on a human pattern which is inappropriate. But if they were not doing this about the basic everyday feelings — about whether their elephant is pleased, annoyed, frightened, excited, tired, sore, suspicious, or angry — they would not only be out of business, they also would often simply be dead.

Inappropriate anthropomorphism is always a danger, since it is easy to get lazy and presume that the way we see and experience the world must be the only way. It is also easy to become self-serving and hope that because we want or need animals to be happy or unfeeling, they are. In fact, the only guard against the inappropriate use of anthropomorphism is knowledge, or the detailed study of the minds and emotions of animals.

I'm happy to report that because of the activism to free her, Ruby was transferred in 2007 from the Los Angeles Zoo to the wonderful Performing Animal Welfare Society elephant sanctuary in Galt, California. She seemed happier in the sanctuary, where she became the "matriarch" of the other three elephants, until she died in 2011.

Ultimately, if we decide against using anthropomorphic language, we might as well pack up and go home because we have no alternatives. Should we talk about animals as a bunch of hormones, neurons, reflexes, and muscles absent any context for what they're doing and why? Anthropomorphism is inevitable and involuntary. Psychologist Gordon Burghardt notes that denying our own intuitions about an animal's experience is "sterile and dull." If we don't anthropomorphize, we lose important information. Maybe it would help if we developed a new term for appropriate anthropomorphism. Burghardt calls for "critical anthropomorphism," while I call for "biocentric anthropomorphism."

The renowned and influential experimental psychologist Donald Hebb, who loved to collect numbers and do statistical analyses, also made some important observations about anthropomorphism. For Hebb, zookeepers' anthropomorphic accounts proved to be "an intuitive and practical guide to behavior," enabling them to best interact with their captive chimpanzees. Hebb suggested that an objective analysis of the basis of anthropomorphism might make it "suitable for the purposes of a scientific comparative psychology."

The Benefits of Anthropomorphism

As this chapter makes clear, many people discuss anthropomorphism as a problem to overcome, as if it were an obstacle to accurately understanding nonhuman animals. But what if, not only is it an inherent aspect of sentience, empathy, and intuition, it also improves human-animal relationships and caretaking?

A very interesting and important 2019 study showed that there is an upside to anthropomorphism because it favors coexistence with other animals rather than domination, and it challenges traditional methods of wildlife management that often cause harm and death. In their research essay published in *Biological Conservation* titled "How Anthropomorphism Is Changing the Social Context of Modern Wildlife Conservation," Michael Manfredo and four colleagues describe how anthropomorphism can influence people's values and conservation practices.

In the research project, 43,949 US adults were interviewed and asked questions, according to an article about the study, "designed to gauge respondents' level of anthropomorphism, such as whether they believe other animals have intentions, emotions, and minds of their own." In order to gauge whether they leaned toward mutualism or domination, people were also asked if they agreed with statements like "Humans should manage fish and wildlife populations so that humans benefit," and "I view all living things as part of one big family."

Around 67 percent of study participants were OK with the use of anthropomorphism; people in Hawaii were the most anthropomorphic (74.5 percent) and people in South Dakota were the least (54.5 percent). In addition, accepting anthropomorphism and favoring mutualism were highly correlated.

This research indicates a shift in people's values in the ways they relate to nature and wildlife management. Traditionally, when it comes to "solving" human-animal conflicts, wildlife management policies favor killing animals, and they encourage hunting, trapping, and fishing, but as the article about the study stated, "People who empathize with animals often disagree with these practices but rarely have a voice in government wildlife agencies." Further, researchers found that "income, education, and urbanization were only weakly correlated with anthropomorphism at the individual level."

In their study, Manfredo and his colleagues "view anthropomorphism as an important cognitive characteristic that affects human thought and behavior, and an adaptive response to shifting social conditions." They conclude, "Anthropomorphism, through its effect in stimulating value shift, leads to challenges of traditional approaches to wildlife management. It emphasizes consideration of *individual* animals and the avoidance of lethal control techniques such as is proposed in the concept of compassionate conservation.

Further research in other modernized countries with similar cultural characteristics is needed to establish the broader generalizability of our findings."

In essence, anthropomorphism favors mutualism rather than an ideology of domination in our interactions with wildlife. Compassionate conservation stresses that the life of each *individual* matters simply because they are alive, not because of what they can do for us. One of its guiding principles is "First do no harm."

Similarly, in a 2012 essay titled "Anthropomorphism as a Conservation Tool," Alvin Chan writes, "Anthropomorphism has the potential to aid conservation biologists to conserve target species by developing empathy among the public, effectively promoting considerate practice."

In 1978, in his book *Thinking Animals*, naturalist Paul Shepard wrote, "Animals embody every quality found in the human personality. In the whole range of human temperament and character there is nothing unique, nothing not found as some aspect of another species. It is the only other place they are found."

This anthropomorphic truth is now increasingly accepted, and it's leading to practical changes in how we relate to nonhuman animals. In March 2023, during a celebration to launch the new Institute for Animal Sentience and Protection at the University of Denver, Colorado's first gentleman Marlon Reis wrote: "We know that nonhuman animals lead rich emotional lives through our own senses and sensations, but science gives us the proof. They feel joy, sorrow, empathy, even embarrassment."

Not only is biocentric anthropomorphism correct, but it is more useful than we might have expected. The seemingly natural human urge to recognize emotions in animals — far from obscuring the "true" nature of animals — may reflect a very accurate way of knowing. As people accept this, it is changing attitudes and guiding actions. This knowledge is essential for making ethical decisions on behalf of nonhuman animals in all sorts of situations in which they are used, abused, harmed, and killed.

{ 6 }

Ethical Choices: Why Animal Well-Being Matters

Ethics in our Western world has hitherto been largely limited to the relations of man to man. But that is a limited ethics. We need a boundless ethics, which will include the animals also.... The time is coming when people will be amazed that the human race existed so long before it recognized that thoughtless injury to life is incompatible with real ethics. Ethics in its unqualified form extends responsibility to everything that has life.

— Albert Schweitzer, *Memoirs of Childhood and Youth*

Clearly, we know a lot more today about animal emotions than we did seventeen years ago, and yet even then we already knew enough to inspire changes in the way we think about and treat other animals. We need to turn our knowledge into action. We must not simply continue with the status quo — using welfare standards that result in millions of animals being mistreated and killed — because that is what we've always done.

What we know has changed, and so should our relationships with animals. We must reflect on the fact that many animals experience deep passion and deep suffering — they feel love and pain — and then consider all the ways we currently treat animals in our society and work to change what is wrong. This is what it means to bridge the "knowledge translation gap," which Jessica Pierce and I discuss in our book *The Animals' Agenda*. The knowledge translation gap refers to how, despite the fact that we know that nonhuman animals are sentient beings, our society's laws and regulations still allow people to cause unnecessary, avoidable, and intentional harm and suffering. This chapter explores several arenas where this is the case — in scientific research, in agriculture, in sports and entertainment, in zoos, and in wildlife conservation — but this list is by no means exhaustive.

Ultimately, we need to look at every place where people and animals intersect and decide if we are caring for animals properly. Far too often, we are not. This impacts each of us individually and all of us as a society, and as a society there are many reasons we should be ashamed of how we interact with other animals. We need more checks and balances in our interactions. We disregard and silence sentience routinely.

When I give talks, some people become defensive and point out all the ways that humans do care for animals. And it's true; many people go to great lengths to treat, nurture, and rescue animals, and there are some wonderful stories to tell about this. I've provided many in this book. However, I stress that very often what passes for "good welfare" in our society simply isn't good enough, and instead we need to focus on animal well-being (see more below). Humans often make distinctions between how they treat "smart" animals and "dumb" animals, and particularly in institutional settings, animal care usually deteriorates significantly as it becomes less convenient (that is, less profitable and/or an impediment to "progress"). This isn't good enough. We must provide the best care we can to all animals all the time and, ideally, work toward not using animals at all.

Our thinking regarding animals and science has changed a great deal in the last century. It was only a hundred years ago that writers and naturalists who gave animals "personalities" were vilified as "nature fakers," a movement that was initiated by John Burroughs and strongly supported by Teddy Roosevelt. Their targets were naturalists such as Ernest Thompson Seton, William J. Long, and Jack London. Science was supposed to be objective and value-free, and mixing "facts" with ethics, values, and emotions was taboo, so "sham naturalists" like London and others were charged with sentimentalizing science. Their nature writing was considered fictional.

That was then. Now we know better. We know that "objective, value-free science" itself reflects a particular set of values. We know that the results of scientific research (all those facts) should influence how we act in the world; otherwise, science becomes a meaningless exercise. We also know animals feel emotions and suffer at our hands, and they do so globally. Ethics, with a capital E, needs to have a place in our ongoing deliberations about how we interact with other animals. And I mean ethics the way Socrates did, as the notion of "how we ought to live." Ethics requires a critical appraisal of who we are and what we do, as well as a vision of what we want to be. Ethics helps us evaluate the best course of action when there are several options and our information is incomplete, conditional, or conflicting. We may never reach our ideal vision, but it's necessary to formulate one so that we have guidance to make the best choices.

When it comes to our relationships with animals, our vision of who they are and what they mean to us requires that we change the way we have always treated them. We know that animals are not "things" that exist for our convenience — they are *beings* who deserve high-quality lives. Nonhuman animals have feelings and thoughts, related and similar to our own, and they deserve respect and consideration. We don't have the right to subdue or dominate them for our selfish gain — to make our lives better by making animals' lives worse. Further, as self-conscious, sentient beings ourselves, we are able to recognize suffering, and we are obliged to reduce it whenever we can. By making decisions that help animals, we add compassion, not cruelty, to a "wounded world," as the ecologist Paul Ehrlich calls it.

There is much that we can do, and much that is relatively easy to do. But before exploring a few of those positive actions, I want to say a few more things about doubt and action.

The Precautionary Principle

The nature of science is that it never (well, hardly ever) yields answers that are complete and unequivocal, but the consensus among scientists is that most, if not all the animals that we use for our own purposes, whether for food, for fun, or for scientific procedures, are sentient. The simplest definition of animal sentience is "Feelings that matter."

— JOHN WEBSTER, "ANIMAL SENTIENCE AND ANIMAL WELFARE"

Because uncertainty never disappears, decisions about the future, big and small, must always be made in the absence of certainty. Waiting until uncertainty is eliminated is an implicit endorsement of the status quo, and often an excuse for maintaining it.... Uncertainty, far from being a barrier to progress, is actually a strong stimulus for, and an important ingredient of, creativity.

— HENRY POLLACK, UNCERTAIN SCIENCE...UNCERTAIN WORLD

Henry Pollack was speaking about attitudes toward climate change, but his remarks apply equally well to the study of animal emotions and animal sentience. Pollack was describing what many scientists call the precautionary principle, which guides the actions of people around the world every day, though they are not always aware of using it. Basically, this principle maintains that a lack of certainty should not be an excuse to delay taking action. Sometimes we have to act based on our best judgment because we may never have "all" the facts, and if we wait for absolute certainly, we might never do anything.

With respect to animal emotions, we may never know everything that goes through an animal's mind and heart, but we know as much as we need to in order to change, right now, the way we treat animals. As a society, we just need to ask ourselves what would cause more harm and what has higher consequences: Treating mammals, birds, fish, and reptiles as if they possessed the full spectrum of emotions and sentience only to one day discover that animals possess only some of these qualities? Or continuing to abuse all animals only to one day discover that every species possesses sentience and an emotional richness equal to humans? Guinea pigs, elephants, dogs, and countless other animals would surely appreciate our use of the precautionary principle.

However, for some people, acknowledging that animals possess emotions is unnerving because of how society currently treats them. What we don't know, or supposedly don't know, makes it easier to avoid feeling guilt or remorse over the inhumane treatment many animals are subjected to. This is true for the scientists I talk about in chapter 5, and it is also true for average folks, zoo administrators, and food company employees. But when the basis for one's doubts is eroded, one is left with only denial — by not acknowledging the rich emotional lives of animals, we can avoid invoking the precautionary principle and changing how we treat them.

Ideally, most people, scientists included, would act only when they're sure. We love to be in control. But certainty is difficult to find in this complex world. We don't really know much about the causes of the common cold or whether or not living near power lines causes health problems, but we try not to take unnecessary risks. As a society, we are unnecessarily risking the lives and welfare of millions of animals every day. We do this, in part, because animals are integral to so many essential aspects of our lives, from the food on our tables to the medicines that heal us, and we see no other way. Solutions that account for animal emotions seem too complex to envision, and they seem like they would entail too many sacrifices on our part. But we need to do something besides deny the problem. As Henry Pollack encourages us, our doubts and concerns should become the fuel for our creativity. Surely, there must be ways we can care for animals and meet society's needs at the same time. Even if we can't always achieve a perfect balance, we should certainly explore every possible option we can think of. Animals deserve at least that much consideration, and our ethics should demand it.

The title of one of Frans de Waal's books is *Are We Smart Enough to Know How Smart Animals Are?* What I wonder is, are we empathetic and compassionate enough to care how emotional animals are?

A Paradigm Shift: Valuing the Life of Every Individual

The paradigm shift that I hope to help inspire is replacing the science of animal welfare with the science of animal well-being. Animal welfare fails countless animals because it allows for horrific abuse as long as we're trying as hard as possible to reduce their pain and suffering. On the other hand, animal well-being stresses that the life of every single *individual* matters, and some of the ways in which we abuse other animals are unacceptable and should no longer continue.

To me, sentience — the ability to feel — is the central reason to better care for animals, and not what animals know, or how intelligent they are. Of course, the fact that an individual is alive is reason enough not to harm or kill them. Does it really matter if monkeys in a zoo, rats in a lab, or cows on a farm ever understand what is going on around them, or what is being done to them by humans, if they can feel pain and experience suffering? Animals in these situations depend on us completely, and their behavior tells us when they're healthy and happy or in pain and sad. Animals can't call 911 in an emergency; they depend on our goodwill and mercy. They surely protest publicly when they're suffering. Their pain is easy to see and all too often ignored. This is what standards of well-being focus on, adjusting our care based on what an individual animal feels.

As practiced today, animal welfare is not much concerned with the plight of *individual* animals, and in numerous instances the welfarist approach patronizes animals, since human interests invariably trump animal interests. Business as usual for welfarists basically boils down to trying to give animals a marginally better life as they're being ruthlessly exploited, abused, and often killed, while leaving unquestioned the human practices that are causing tremendous animal suffering. Welfarism is a salve for our conscience, but without enacting any meaningful protection or care for animals.

Animal welfare science is going strong and an internationally recognized field of research. But where exactly is it headed? On the one hand, there have been some positive changes on behalf of animals. In March 2016, China released its first set of guidelines for the more humane treatment of laboratory animals, and Iran banned the use of wild animals in circuses. These are positive moves, but they are not enough. Much more often, welfare science weaves a cloak of objectivity around abusive practices. Donald Broom and David Fraser write, for example, that "the assessment of welfare can be carried out in an objective way that is independent of any moral considerations." This is precisely the problem: defining *objectivity* as separate from ethics, and focusing on the welfare of species and not the well-being of individual animals. This simply becomes a quick way to ignore the *subjective* lives of the animals who are being used and abused.

This is precisely why, in *The Animals' Agenda*, Jessica and I developed the idea of the *science of animal well-being*, which focuses on individuals — their own personal joys and sufferings — rather than on species. If the well-being of individual animals was prioritized, this would not allow animals to be used and abused in the ways that welfarism allows. Welfarism puts human needs first and, without sacrificing those needs or doing so far too infrequently, tries to accommodate animals. Well-being broadens the question by asking what do animals themselves want and need. It goes beyond the welfare box and tries to understand animal preferences from the animals' point of view.

For example, welfarism asks whether mink on a fur farm would prefer taller or shorter cages; well-being challenges the idea that mink should be in battery cages on a fur farm in the first place. Animals cannot have true well-being or "good lives" under such conditions — no matter how many welfare modifications are made to the cages. Simply put, well-being sets a considerably higher bar than welfarism because it requires humans to change in order to accommodate what animals want and need.

Moving Beyond the Five Freedoms

All animals have the right to live the lives they're supposed to live — to express who they are and perform natural behaviors absent human domination and control. In 1965, the British Farm Animal Welfare Council created the "Five Freedoms of Animal Welfare," and these have since been adopted by developed nations as the basis for their treatment of animals. The Five Freedoms are as follows: freedom from hunger and thirst, freedom from discomfort, freedom to express normal behavior, freedom from fear and distress, and freedom from pain, injury, and disease. Jessica Pierce and I refer to the Five Freedoms as "fake freedoms" because, although they might improve the lives of some animals, many if not most animals who are used for food and other reasons don't really have good lives and are treated as if they are ultimately disposable. The Five Freedoms basically help make people feel good about the harm for which they're responsible.

Welfare concerns generally focus on preventing or relieving suffering, and making sure animals are being well-fed and cared for, without questioning the underlying conditions of incarceration and abuse that shape the very nature of their lives. We offer lip service to freedom when talking about "cage-free chickens" and "naturalistic zoo enclosures." These are simply incremental improvements to animal prisons. Real freedom for animals requires a deep examination of our own behavior, which we avoid. Real freedom might mean changing the way we treat and relate to animals so animals are truly free of human impacts.

In *The Animals' Agenda*, Jessica Pierce and I add a sixth freedom — freedom from extinction. We also ought to account for our impacts on wild species, even when those impacts are indirect or not specifically targeting certain wild species. The sixth freedom is really the freedom of animals to live in habitats in which survival is possible — not merely eking out a poor existence, but thriving, even if their lives are not as free as they would like nor as free as their ancestors.

Humans' relationships with animals are also part of the cultural heritage of many communities, and we are altering these centuries-old patterns. In *Where Have All the Animals Gone?* by Dale Peterson, photographer Karl Ammann tells the story of one of the guides at a camp in Samburu in northern Kenya. The guide told them that when he was a boy he used to herd cattle and see giraffes every day. Yet his children have never seen a giraffe where they live.

In *Unleashing Your Dog*, Jessica and I expand the Five Freedoms into ten freedoms and show how they can be applied to dogs and other companion animals. The expanded set reads: freedom from hunger and thirst, freedom from pain, freedom from discomfort, freedom from fear and distress, freedom from avoidable or treatable illness and disability, freedom to be themselves, freedom to express normal behavior, freedom to exercise choice and control, freedom to frolic and have fun, and freedom to have privacy and "safe zones." These ten freedoms are echoed by Martha C. Nussbaum's "capabilities approach," which she describes in her book *Justice for Animals: Our Collective Responsibility*. Nussbaum argues that, as a matter of justice, animals must be free to perform their characteristic form of life — doing what comes naturally.

Did the crafters of the Five Freedoms fail to recognize the fundamental paradox of their approach to animal welfare? How can an animal in an abattoir or battery cage be free? Being fed and housed by your captor is not freedom; it is simply being kept alive. Indeed, the Five Freedoms and the concept of animal welfare are not really concerned with freedom or welfare per se. Rather, they maintain the status quo for humans while continuing to keep animals under conditions of profound deprivation.

Four "Rules" for Interactions in Nature

In his landmark book *Respect for Nature*, philosopher Paul Taylor proposes four basic ethical principles or "rules" to guide our interactions with wild animals in nature: the rule of nonmaleficence, the rule of noninterference, the rule of fidelity, and the rule of restitutive justice. Again, these are ways to respect animal well-being for the individuals and species who are trying to coexist with us.

The rule of nonmaleficence is "the duty not to do harm to any entity in the natural environment that has a good of its own." This involves a negative duty, or not taking actions "that *prevent* harm from coming to an entity or those that help to *alleviate*...suffering." In essence, through our actions, we should not destroy their status as wild animals.

The rule of noninterference states that humans have a duty "to let wild creatures live out their lives in freedom." In essence, our intrusions into "the domain of the natural world...terminates an organism's existence as a wild creature." Once we interfere in animals' lives, we become part of their existence, and no matter how heart-wrenching it is to do nothing, we should not interfere in natural systems. As a fieldworker, I can't put into words how difficult it is to sit by, leave animals be, and let nature take its course when horrible things are happening that we could stop.

The rule of fidelity means we should not break a trust that a wild animal places in us. This trust is shown by their behavior, and it means not deceiving them by misusing this trust. Some fieldworkers do indeed believe that the animals they study come to trust them. For example, Jane Goodall claims that her relationship with the chimpanzees she intensively studied "can best be described as one of mutual trust."

Taylor's rule of fidelity also can be applied to our interactions with domesticated animals and other individuals. In fact, we have at least equally compelling reasons not to break the trust of companion animals, since they depend on us and form close relationships with us. Lawrence Johnson notes:

"Certainly it seems like a dirty double-cross to enter into a relationship of trust and affection with any creature that can enter into such a relationship, and then to be a party to its premeditated and premature destruction." Certainly, when a person breaks another person's trust in ways that cause harm we consider that immoral.

Finally, the rule of restitutive justice "imposes the duty to restore the balance of justice between a moral agent and a moral subject when the subject has been wronged by the agent." Simply put, if we cause harm to animals, we must make amends and try to fix the problem or end the harm. If we followed that rule, how might that impact and change human-animal relations in the following arenas?

In the Lab: Take a Number

As a scientist myself, I tend to get the most upset about animal abuse in the lab. It's probably unrealistic and self-centered, but I think that scientists should hold themselves to higher standards than nonscientists. I feel this way not just because the actions of one thoughtless scientist reflect badly on all scientists, but also because the rest of society often takes their cues from us. If scientists, who should know better, feel free to treat animals like disposable objects, then that gives permission to others to do the same. However, were scientists to insist on conducting only humane research because it is the right thing to do, then it might be much harder for others to justify animal abuse.

Around 100 million animals are used in laboratory research globally every year. In the United States, it's estimated that about 800,000 warm-blooded animals are used each year, including 112,595 cats, 71,921 monkeys, and 44,847 dogs. However, USDA estimates don't include rats, mice, and birds, the animal who are used the most. In a detailed analysis of animal use in 2017–18, veterinarian Larry Carbone estimated that 111.5 million rats and mice were used each year, and about 44.5 million were used in potentially painful experiments. According to Cruelty Free International, the number of animals used in experiments in the United States rose by 6 percent in 2021. They also note that in 2021, 70,161 animals were used in experiments in which no relief is provided for animals experiencing pain or distress, called category E experiments. The most frequent victims of these painful experiments were hamsters and guinea pigs, but in 2021, 2,583 rabbits, 1,621 monkeys, and 360 dogs were also used in horrific, category E tests.

These numbers are rough estimates. While they are very small numbers compared to the billions of animals used in the commercial food industry, this is still a lot of animals. The vast majority of these animals live most of their lives in extremely small cages, bored and lonely, and they usually die in the lab when researchers are done with them: Almost all are killed by researchers — either deliberately, since dissecting them is part of the research and sacrificing them clears the way for new subjects, or as an unintended consequence because the animals just couldn't cope with what was being done to them.

The scientific community has created a set of professional standards that is supposed to guide scientists to preserve animal welfare as much as possible in their research. In theory, animals used in research in the United States are protected by the Animal Welfare Act (AWA). But so far it has proven inadequate. Only about 1 percent of animals used in research in the United States are protected by this legislation, and the legislation is sometimes amended in nonsensical ways to accommodate the "needs" of researchers. For instance, here is a quote from the 2004 federal register: "We are amending the Animal Welfare Act (AWA) regulations to reflect an amendment to the Act's definition of the term *animal*. The Farm Security and Rural Investment Act of 2002 amended the definition of *animal* to specifically exclude birds, rats of the genus *Rattus*, and mice of the genus *Mus*, bred for use in research."

Of course, for both science and common sense, this is nonsensical. Birds, rats, and mice are animals by every other definition. However, since researchers are not "allowed" to abuse animals under AWA, the definition of *animal* was simply revised so that it refers only to creatures whom researchers use less frequently, even if the excluded animals are also sentient and empathetic beings.

Horror Stories: From the Lab to the Island of Dr. Moreau

Tales of extreme abuse grow increasingly rare, and I truly believe that many scientists sincerely try to do their best for the animals in their care. But even "everyday" stories of lab work can sometimes make me feel sick and woefully sad. For instance, the Animal Welfare Act legally allows chimpanzees to be kept in cages too small for them to move as much as they need to. This legislation certainly aids labs by saving space, but where is the welfare for the chimps? In response to the discovery that mice are empathic rodents, it was suggested that an opaque barrier be used to separate mice so that they can't know what's happening to another mouse. This is because mice who observe each other during experiments may change their behavior as a consequence and therefore "contaminate" the data. I'm sure if you saw what was going on in the other cage, you'd prefer not to see it either.

Even scientists who see what's going on can sometimes have trouble breaking through the veneer of their own jargon-garbled "objectivity" and telling it like it is. To read their convoluted explanations is to feel like you've entered the theater of the absurd. I once read a report about pain in pigs that concluded:

> The observed changes of acoustical parameters during the surgical period can be interpreted as vocal indicators for experienced pain and suffering. We conclude that a careful analysis of the vocal behavior of animals may help to gain a deeper knowledge of pain, stress, and discomfort that an animal perceives. The results deliver further facts for a critical reevaluation of the current practice of non-anaesthetized castration of piglets.

What is this saying? That castrating young pigs without anesthesia, a routine procedure in domestic pig production, is painful. The piglets don't like it, as evidenced by their squeals and attempts to get out of their horrible situation. The researchers conclude that it can be "interpreted" — a far cry from concluding the obvious — that the screams of animals really mean something after all.

Some scientists willingly push the legal boundaries of what's permissible, no matter how much pain and suffering they cause. For instance, using a $500,000 taxpayer-funded grant provided by the US Department of Justice, a University of Wisconsin professor attached electrodes directly to the skin of pigs and implanted catheters near their hearts to measure the effects of Taser shocks. In this way, live pigs were electrocuted to death in order to satisfy claims that Tasers are safe. Then there were psychologist Harry Harlow's well-known maternal deprivation studies in monkeys, in which the good doctor decided to study the nature of mother-infant bonds by taking infant monkeys away from their mothers to see what would happen. Is there any question in anyone's mind what happened? More to the point, were the questions these researchers seeking to answer so desperately vital that they justified suffering and even death?

I believe there are times when trading new knowledge for fewer traumas is a worthwhile exchange. But what's especially disturbing is that often the knowledge gained in animal research isn't even used to make human or animal lives better — because the people who are in the position to implement the new findings are unaware they even exist. Psychologist Kenneth Shapiro, in his book *Animal Models of Human Psychology*, shows this to be true in the case of animal models of human eating disorders.

As well, species differences often preclude the utility of animal models for helping humans, a point well made by, among others, physician Ray Greek and his veterinarian wife, Jean. In February 2006, the prestigious Diabetes Research Institute published a report stating that "scientists from the Diabetes Research Institute at the University of Miami Miller School of Medicine have shown that the composition of a human islet is so different than that of the rodent model, it is no longer relevant for human studies." And in an essay published in the *Journal of the American Medical Association*, it's been estimated that 106,000 people die each year in hospitals from adverse reactions to drugs that had previously been tested on animals and were approved by the Food and Drug Administration. Adverse drug reaction is the fifth-leading cause of death in the United States, after heart disease, cancer, stroke, and lung disease.

In 2023, it was reported that the unnatural and sterile lives of lab mice can undermine science by providing data that have no bearing on human disease. Of course, this begs the question whether any mice should be used at all. As a matter of standard practice, many studies exclude female lab mice because they're believed to be less predictable than males due to their fluctuating hormones. Yet even this has been disproved. It turns out male mice, usually the default choice, actually behave more erratically than females.

In a 2023 essay called "Could the Next Blockbuster Drug Be Lab-Rat Free?," Emily Anthes writes, "Momentum is building for nonanimal approaches, which could ultimately help speed drug development, improve patient outcomes, and reduce the burdens borne by lab animals, experts said." As and when this happens, mice and other lab animals who aren't protected by the US Animal Welfare Act will benefit, as will the millions of people who rely on a wide variety of drugs to help them along.

Finally, there are scientists who, if they think they can get away with it, ignore any legal and ethical boundaries to conduct their work. For years, there were unsubstantiated allegations that the Oregon Regional Primate Research Center (ORPRC) was violating federal animal welfare laws. In 1998 Matt Rossell took a job there to find out. During the course of his undercover investigation, he documented a number of outrageous cruelties, such as the basic living conditions: At the time, about a thousand monkeys lived in the laboratory, many of them in tiny and often filthy cages not much larger than four square feet. But by far the most horrific practice Rossell observed was "electro-ejaculation." This involved strapping an unanesthetized adult male monkey into a restraining chair, wrapping two metal bands around the base of his penis, and applying an electrical charge to cause ejaculation. The procedure was performed in order to collect semen samples. The monkey, number 14609, was nicknamed "Jaws" because one of his supervisors taught him to bite the bars of his cage. Until the procedure was banned as a result of Rossell's investigation, Jaws underwent electro-ejaculation on 241 separate occasions from 1991 to 2000, not counting the times when his penis was shocked more than once during a session to get the semen needed. No wonder Jaws tried so desperately to escape the ordeal. In the end, a veterinarian resigned and some of the scientists working there made critical statements about conditions in the laboratory, but the center continues to operate and is now called the Oregon National Primate Research Center.

Professional Standards: The Three Rs

Among scientists, it's commonly accepted that there aren't any substitutes for animal research. They say it's necessary for the development of vaccines and medicines, to help us understand biological and cognitive processes, and of course for the development of commercial products, from dyes to Tasers, to make sure they don't harm humans. Yes, some researchers think they should be able to do whatever they want to the animals with whom they work, but the great majority try to minimize the suffering their research causes. Of course, many procedures that seem like "necessities" today merely demonstrate a lack of creativity; as we'll see, when researchers really emphasize animal welfare, they're finding noninvasive ways to get the information they need.

In any case, the long-standing professional guidelines that most scientists follow are called "the Three Rs": This refers to refining procedures that cause harm to animals, reducing the number of animals used, and replacing animals with other methods whenever possible. The term "the Three Rs" was proposed by William Russell and Rex Burch in their 1959 book *The Principles of Humane Experimental Technique*. The Three Rs call attention to the fact that animal welfare is important. Practicing scientists are not supposed to merely give lip service to this. They are expected to do all they can to minimize the impact of research on animals, or to eliminate such research if they can.

The Three Rs are guidelines, however, not laws or an official code of conduct. No one reviews a scientist's planned research and suggests alternatives. Each individual decides how to follow the Three Rs, and in practice, the Three

Rs are often sacrificed in the name of expediency. They don't stop researchers from conducting highly invasive experiments that harm and kill their unfortunate subjects. In the rush to find a treatment or a cure for this or that disease, it's easy for scientists to justify more animal research, using invasive techniques they are familiar with (and that often kill the animal), rather than think up more humane alternatives. And they do this despite the fact that animal testing provides no guarantees — huge numbers of people continue to suffer or die even after drugs or other treatments have been tested "successfully" on animals. In practice, the false hope of animal research is an enormous waste of money and life.

In fact, there is a plethora of data that shows that animals who are treated humanely and who are not stressed produce "better data." This doesn't justify animal research, but it does call attention to the fact that if animals aren't properly cared for, much of the information that is collected is compromised and even useless. If scientists actually upheld the Three Rs, not just when it was convenient but at all times, it would make for better science, better data, and perhaps a slightly more compassionate world. However, I agree with neurologist Dr. Aysha Akhtar who stresses, "What we really owe the animals is to legitimately replace their use." We also can make contributions to organizations that support finding alternatives to using animals or who provide care to retired lab animals.

Some say life is too short to ponder if animals have feelings, and I agree. However, I also feel that life is too long to ponder whether animals have emotions. The suffering that numerous animals endure while some researchers still try to figure out if they're feeling anything at all is unacceptable and cruel.

Bad Fieldwork: Harming Animals in Their Homes

Sound science requires animal subjects to be physically, physiologically, and behaviorally unharmed. Accordingly, publication of methods that contravene animal welfare principles risks perpetuating inhumane approaches and bad science.

— KATE FIELD, "PUBLICATION REFORM TO SAFEGUARD WILDLIFE FROM RESEARCHER HARM"

There is nothing more exciting than watching animals in the field, their natural homes. The data that are collected are "more real," as one colleague puts it. It reflects what the animals do and how they live in their natural homes or when they're free to move about when and where and with whom they want. But fieldwork, like research on captive animals, isn't always user-friendly to the animals themselves, a concern many researchers and I have had for a long time. It seems silly to have to remind people that wild animals have rich and deep emotional lives, but often this fact is ignored or acknowledged and pushed aside when a study is planned and carried out. The feelings of wild animals matter as much as those of companion and other animals.

It's also important to point out that when researchers negatively impact the animals they are studying, that influences what the animals do, so the quality of the data collected is compromised and the data are misleading. Picking up a mouse by their tail, something numerous researchers do, can be very traumatic, and isolating an animal can terrorize them. In 2022, Australian bird expert Dr. Gisela Kaplan published an outstanding review of what the results of fieldwork on animals means and how animals can be influenced by researchers. Kaplan writes:

> Worse still: human-induced fear can lead to natural top predators increasing their kill rate. Invasive and aggressive interventions by humans can also lead to the collapse of entire niche structures and species coexistence. The point is that the instigators of field research that may have brought about any such harm, even if only just going to nest sites and taking some eggs, then walk away and do not see what behaviour follows. The data then collected, however, reflect the post-manipulation phase of behaviour, and it may not be at all what an individual, pair, or group might have done without such intervention. In summary, there is plenty of scientific evidence now that we, as a species and as researchers, have had a detrimental effect on birds and on wild animals in general.

Kaplan also notes, "In personal interactions, there have been ecologists who became extremely aggressive and angry and dismissed cognition and emotions as 'rubbish,' and on the topic of conservation they asserted that academics dealing with cognitive ethology or compassionate conservation 'didn't know what they were talking about,' while others, in personal contact, felt bad about causing pain to animals but accepted this, if it was for 'the greater good.'"

The researchers' arrogance and dismissal of the emotional lives of the animals they are studying is reprehensible. How researchers influence the behavior and emotions of animals under study — what is obvious and what is hidden — is among the reasons why some people argue that noninvasive research produces more accurate and useful data. The information these and other researchers are collecting and analyzing isn't representative of who these birds are. This bad and unreliable data is similar to the information that is collected in invasive laboratory research. I often have suggested that when people talk about and write the results of these studies, the study titles should read something like "The Social Behavior of *Stressed* Birds." It is their responsibility to tell listeners and readers that they *know* they have toyed with the emotional lives of the animals and their results represent birds who aren't typical of their species, but rather birds who have been negatively treated and studied under imposed conditions.

This admission might help people appreciate what I call "the biodiversity of sentience" — when researchers admit they've caused pain and suffering, that would alert people that the animals are sentient, feeling beings. More to the point, researchers doing fieldwork must take all precautions to be sure they are not influencing the emotions of the animals they are studying. When data are contaminated, we let the animals down, and what we learn might not be helpful to anyone, since it paints a false image of who animals are and what they do.

Bad Data and Captivity Effects:
The Effects of Cages and Isolation

Much of the research on animals is conducted in captivity where animals are often stressed. Stressed animals don't behave like unstressed individuals, and particularly when it comes to animal emotions and behavior, information that's collected on anxious or bored animals doesn't have much to say about the normal behavior of individuals of the same species. The term *captivity effects* refers to the range of physical, physiological, and even neurobiological changes induced by captive conditions, and they are rampant among nonhuman animals.

Like us, all animals want and need food, water, air, and sleep. They also need shelter and safety. Like us, they have what might be called "higher-order" needs, such as the need to exercise control over their lives and their environment, to make choices and be creative, to form meaningful relationships, and to engage in their natural behaviors, like play. Some measure of freedom is fundamental to satisfying these higher-order needs in order to thrive and be happy. Animals value their freedom like we do, and they suffer captivity effects that are very similar to ones that we do.

Captivity leads to measurable physiological changes in the brain, including loss of neural plasticity, long-term activation of the hypothalamic-pituitary-adrenal axis, and permanent changes in brain morphology. The stresses of captivity can damage the central nervous system and lead to changes in immune function, reproductive behaviors, and circadian rhythms. Like psychological trauma, the loss of freedom often manifests in observable abnormal behaviors called "stereotypic behaviors," or "stereotypies," that are invariant, repetitive, and serve no obvious function. I simply regard this as "captivity-induced madness." Captivity drives animals crazy. Further, these stereotypic behaviors do not occur in the wild; they are only a product of captivity.

Scientist Françoise Wemelsfelder notes that researchers often use the neutral term *boredom* to describe the abnormal behavior of animals who are permanently housed in small barren cages. Wemelsfelder defines boredom as "the impaired ability to actively focus attention upon, and interact with, the environment." While genuinely bored animals spend more time sitting, lying down, and sleeping than do more-stimulated individuals, captivity-induced stress is recognized by overreactions to novel or unexpected events with increased fear and aggression and by a range of repetitive stereotypies, such as pacing back and forth, bar biting, object licking, tongue rolling, excessive drinking, and much more. To quote Wemelsfelder:

> Sometimes such behavior can be damaging to other animals; licking and nibbling tails and ears of offspring may for example induce cannibalism in rats and mice. As time of confinement proceeds, such patterns tend to become increasingly directed toward the animal's own body or products thereof. Primates may spend long periods of time masturbating, rocking their own body, or eating and regurgitating their own feces. Rats may chase their own tail, tethered sows may show long bouts of chewing air, with no other apparent effect than producing large amounts of saliva. Such tendencies may eventually develop into various forms of compulsive self-mutilation. Laboratory monkeys gnaw at their own limbs or genitals, while parrots will pull out their feathers until completely naked. In summary, the overall decrease in interaction shown by captive animals comes to expression in a decrease in behavioral variability and an increase in self-directed behaviors.

These behavior patterns may sound "abnormal," but they're actually normal responses to being kept in barren cages with few if any social companions or objects. Wemelsfelder notes, "They bear a strong resemblance to behavior pathologies in human beings. It is generally accepted that in human beings, abnormal behavior may be a sign of depression, or other forms of subjective suffering." Thus, one unintended result of most lab work is to further demonstrate that animals have emotions and can experience suffering. In a series of excellent research papers on abnormal repetitive behaviors (ARBs), Georgia Mason and her colleagues have shown that at least ten thousand captive wild animals in zoos worldwide show these stereotypies. As a result of this groundbreaking research, Mason calls for "zero tolerance of such ARBs" because they indicate poor individual welfare and clearly raise serious ethical issues.

Wemelsfelder has studied boredom in animals in great detail and has discovered that boredom can be alleviated using environmental enrichment, which consists of providing toys, straw, bedding, and social companions. Making animals work for their food also has been useful for reducing boredom. Rabbits and other animals react to more stimulating environments by showing increased activity and breeding success and decreased aggression.

"If such measures are combined to enhance the active, flexible character of species-specific behavior," Wemelsfelder argues, "both the well-being of captive animals and the quality of scientific research benefits significantly." In other words, happy animals make for better data. They become a more "normal" model for scientific investigation, and their responses to experimental situations is more, rather than less, stable and consistent. In one Dutch scientific laboratory, environmental enrichment led to a considerable decrease in the number of animals needed for experimentation — precisely because happier animals enhanced the reliability of the experiment's results. This clearly shows caring for the well-being of animals, and seeking to understand their needs and requirements can go hand in hand with enhancing the quality of scientific research. It's also known that enriching the lives of captive animals makes for happier caretakers; it increases the well-being of the people responsible for the care of the animals.

It's also been shown that the richness of a primate's environment affects the animal's brain structure and learning abilities. Professor Charles Gross conducted a study of marmosets that suggests that bored monkeys make for stupid monkeys. In the study, pairs of adult marmosets were housed in one of three types of cages for one month. The first was a bare cage with a food bowl, which actually meets the minimum standards of the National Institutes of Health (NIH). The second was a slightly larger cage with toys and structures, including branches hiding live worms, that encouraged the monkeys to forage. The third cage was twice as big with many more toys.

Unfortunately, the marmosets used in this study were killed so that their brains could be examined. Gross learned that the animals housed in the second two cages developed denser neuron growth and almost double the amount of certain synaptic proteins that the brain uses to relay messages between neurons. He suggests that it may be necessary to house animals being used for cognitive studies in more stimulating environments than those specified by the NIH. Indeed. Even further, now that this particular study has been conducted, whether or not it should have been, there is no reason to repeat it and kill the monkeys we are trying to learn how to care for better.

Noninvasive Research: Better Science through Technology

Traditionally, animal research involving cognition, behavior, and emotion ends with the death of the animal so that the brain can be cut up, blended, or otherwise analyzed. But it seems strange to me that scientists would consider it normal to kill a mind to study a mind. Thankfully, a growing number of scientists agree that it's strange, and they are developing a number of noninvasive research techniques that improve and preserve the lives of the animals being studied, and this, as we've just seen, also improves the quality and reliability of the information collected.

One noninvasive technique of studying other animals is simply closely observing them and filming them when necessary. Of course, as I note above, it's essential to be sure that "just being there and watching them" doesn't change their behavior. Years ago, some researchers in what were called the "hard sciences" — physics, chemistry, lab-bound comparative psychology — often dismissed ethology — simply watching animals — as "soft science," a hangover from natural history and sometimes likened to stamp collecting. Yet any science must provide a description of its domain and it is important to know what animals do if we are to explain why they do it. I can't tell you how many times people who have conducted invasive research have asked me what the behavior they were studying really means! I told them it would be good if they knew that *before* they invaded an individual's body or brain. Remarkably, some agreed, including those who thought a behavior meant one thing only to learn it might mean something else or have multiple functions and explanations depending on context, like aggressive actions.

While research that stresses and harms animals often produces misleading data, which makes it difficult to find answers to the very questions in which we're interested, we can learn a lot from looking at brains at work by using methods that minimally stress the animals being studied. What has been particularly exciting and important is the use of noninvasive neural imaging techniques, such as positron emission tomography, also known as PET scans, and functional magnetic resonance imaging (fMRI). These imaging techniques give researchers the ability to localize the activity of the brain when animals are in a specific social situation. The use of fMRI and PET scans is providing solid neurobiological proof for much of what we suspected all along when it comes to cognition and emotions, confirming, for example, the huge hippocampus in elephants. The field of social neuroscience as it is usually called is rapidly growing and will produce a lot of interesting information in the coming years about empathy, forgiveness, cooperation, mother-infant interactions, and emotions in general.

Here's one exciting example. James Rilling and his colleagues have used PET scans to study the neural response to staged social interactions in dominant male rhesus monkeys. The researchers set up a situation in which nine dominant males were faced with a "challenge situation," in which they witnessed a potential sexual interaction between a rival male and their mate, and a control situation in which only their female consort was present with no potential rival. In preparation for the PET scan, the monkeys were injected with a fluorescent dye and then sedated so that their blood could be drawn and measurements of the brain's uptake of the dye could be assessed; however, the monkeys didn't have to be killed, as usually happens in studies like this.

The researchers were especially interested in male sexual jealousy — they fearlessly used the J word — when males are challenged in having exclusive access to their mates. They discovered similarities between how sexual jealousy is expressed in humans and in these monkeys, and they found there might also be a similar neural network involved in the expression of vigilance or wariness and anxiety shown by a challenged male.

Also, in one study done on living dolphins, Dorian Houser and his colleagues investigated how hearing and echolocation are processed in dolphin brains. By using brain scans, they monitored blood flow in what's called the melon of the dolphin brain, the part that's responsible for processing information about the clicks used in communication. Houser even took pains to reduce anxiety by giving the dolphins Valium when their brains were scanned.

While noninvasive fMRI neuroimaging has been used mainly on people, it holds a lot of promise for studying animal emotions, and there is an increase in the use of this technique on various conscious animals, including dogs, mice, and nonhuman primates.

In another study of jealousy, this one on dogs, fMRI was used to discover what, perhaps, seems obvious: Dogs don't like being dissed. Conducted by Dr. Peter Cook and his colleagues, the study included thirteen dogs who voluntarily partook in this project and who entered into a scanner and remained calm. Their brains were scanned in three different situations: when a dog got food, when a fake dog got food, and when food was deposited in a bucket. The researchers looked at activity of the dogs' amygdala, using aggression as a measure of jealousy, and the results are extremely interesting. Specifically, the awake fMRIs showed a significant relationship between amygdala activity and dog-directed aggression. The researchers wrote, "Aggressive temperament was positively correlated with bilateral amygdala activation when viewing their caregivers providing food to a realistic-looking fake-dog relative to dropping the food in a bucket." The difference in brain activity was a possible indication of jealousy.

There's also a useful potential practical side to this important research. The researchers also discovered that the amygdala response in dogs who displayed aggression habituated with repeated exposures to the same situation. In other words, it's possible that repeated exposures to that specific situation might help to reduce or eliminate dog-dog aggression. They also point out that there might be behavioral correlates of amygdala activation, perhaps facial expressions, body postures, gaits, or vocalizations, and these could be used to predict when dogs are likely to engage in aggressive interactions. Do dogs feel jealousy? Of course. But if this research helps us identify useful predictors of aggression, that might provide a wonderful way to decrease or eliminate these sorts of encounters.

I can only hope that more examples of humane research will lead to better treatment of animals everywhere. We should always ask hard questions and make sure science and scientists are held accountable for their treatment of animals.

Weighing the Benefits of More Research Versus the Costs of Inaction

We've learned a lot about the emotional lives of nonhuman animals and sentience over the past two decades. And while there is still a lot to learn, I find myself frequently asking: Who's really benefiting from more science? Are we? Are animals? There is no doubt that increasing our knowledge will continue to uncover extremely interesting discoveries, but people often ask me, "How much is really new?" To be honest, less and less. Research often confirms what we already suspected or knew, but for which we lacked confirming data. And this brings us back to the precautionary principle: Haven't we learned enough by now to grant animals the right to better lives based on what we already know about their minds and emotions? One prominent academic recently asked me, "Are there diminishing returns for the animals from studies of cognitive ethology, comparative psychology, and philosophy of mind? Don't we need more action on their behalf rather than more data and words?"

My simple answer is yes. From the point of view of diminishing returns, scientists should reconsider and forgo conducting yet more invasive studies of animal emotions, especially when animals are caged and intentionally harmed or killed. Even studies of so-called noninvasive research should carefully consider the costs of the research in terms of individual well-being against the benefits of providing yet more proof about what we already know. Years ago, we could have and should have used what we know to bridge the knowledge translation gap and treat nonhuman animals much better, with more compassion, dignity, respect, kindness, and empathy. Certainly, considering all the research that's been done over the past twenty years, we haven't learned anything that would justify treating nonhumans worse. Proposing less animal research is *not* to be antiscience — it's a plea to use existing science to inspire more protections for other animals. If there ever was a justification that we needed to cause pain in order to learn about what animals feel, that day is long past.

On the Farm: It's Who We Eat, Not What We Eat

If slaughterhouses had glass walls, everyone would be vegetarian.
— Paul McCartney, "'Glass Walls' with Paul McCartney"

While the treatment of animals in other settings can often range from thoughtless to reprehensible, it doesn't compare to the horrific ways in which so-called food animals are treated. So I want to begin with the reminder that animals are beings with feelings, not objects, and when we talk about meat, we should be asking *who*'s for dinner, not *what*'s for dinner.

Many solid exposés of factory farming and the meatpacking industry have been published, so despite the scale of the problem, I will not spend much time cataloging agriculture's sins. However, if ever there were an area where our society should begin practicing the precautionary principle in regards to animal sentience, this is it. If we are what we eat, then who we eat redefines cruelty. In fact, it's impossible to conceptualize how much death occurs in a single day.

By the Numbers

While numerous animals are used for research, education, clothing, amusement, and entertainment, their numbers pale when compared to the numbers of animals used for food.

In 1961, in the United States alone, it's estimated that we slaughtered around 7 billion land animals for food. Today, while estimates vary, we are killing anywhere from 55 to 70 billion land animals per year, and perhaps as many as 23.3 million land animals are killed each day. We also kill around 80 billion farmed fishes every year.

The number is astronomically higher globally, and when wild sentient fishes are included, the number soars to the trillions.

The "animal kill clock" keeps tabs on the number of animals killed annually for food, and the number that shocks me the most is the number of animals killed since I last visited their page. One day, it estimated that more than 22,000 animals were killed for food in the United States every ten seconds.

Here is another detail that helps quantify the carnage. In 2022, slaughter lines for pigs increased in speed to increase profit and production. The US Department of Agriculture (USDA) permitted food processing plants to increase the number of pigs slaughtered to more than 1,106 per hour per plant.

It's also interesting and extremely disturbing to note that the biomass of livestock currently dwarfs that of wild mammals. A sad 2023 report read, "A team headed by Prof. Ron Milo found that the biomass of livestock has reached about 630 million tonnes (metric tons) — 30 times the weight of all wild terrestrial mammals (approximately 20 million tonnes) and 15 times that of wild marine mammals (40 million tonnes)." This means that the weight of all remaining wild land mammals is less than 10 percent of humanity's combined weight, and only 6 percent of all the land mammals on Earth are wild.

The Global Impact on the Environment

Putting ethical issues about extreme animal abuse aside for the moment, there are other reasons to consider opting out of the meat economy. One major one is environmental, and the problem is much worse today than when this book was first published seventeen years ago. The feedlots and slaughterhouses of factory farms are responsible for enormous environmental degradation and losses in biodiversity.

According to Lucas Reijinders and Sam Soret, compared to soy production, meat production takes more land (six to seventeen times as much), water (four to twenty-six times), fossil fuels (six to twenty times), and biocides (six times as many pesticides and chemicals used in processing). Though it sounds funny, cows are also a significant source of methane. To quote James Bartholomew: "A single dairy cow belches and farts 114 kilos of methane a year. It is a methane machine. Methane is far more lethal as a greenhouse gas...than carbon dioxide. It is 23 times more potent, although it does not last so long in the atmosphere. The methane produced by a single cow is equivalent to 2,622 kilos of carbon dioxide." In 2023, it was estimated that the total population of cows globally — around one billion — is responsible for 220 trillion pounds of methane a year. Furthermore, according to Philip Fradkin, the feeding of livestock is still the single greatest use of the dwindling amount of water in the western United States.

One report said, "Food production accounts for over a quarter of all global greenhouse gas emissions." This means that cows and "food animals" alone are causing major harm to Earth. But when considering the overall impacts of agriculture and food production, human impacts on our magnificent planet are astonishing. Since 1970, human activities have wiped out 60 percent of wildlife populations, and we have lost half of the topsoil on the planet in the last 150 years and are losing about 26.5 billion tons every year.

Life in the sea is also being destroyed by human activities through ocean acidification and fishing. We kill between 1 and 3 trillion wild aquatic animals every year for food. It is estimated that if we keep fishing at the current pace the oceans will be empty of fish by 2048.

Cruelty in the Slaughterhouse

Some farmers resent the use of the phrase *factory farm* for today's modern industrial slaughterhouses because it makes their operations sound like what they are, factories for turning sentient animals into meat. To me, the term *factory farm* isn't strong enough for the reprehensible hell holes these places are for their residents. The worst slums in the human world do not compare to the ways many sentient cows, chickens, turkeys, and pigs spend their entire lives: crowded into tiny cages and small fenced lots; standing, eating, and sleeping in mud and their own feces; and having their lives ended by a mechanical butcher. In fact, countless chickens, cows, and pigs die of stress, injury, or disease before they are mature enough for the slaughterhouse — and some may die of broken hearts as they realize what is happening to them and the horrific, painful ways they are about to be killed.

What happens in slaughterhouses is nauseating and thoroughly inhumane, and it includes electrocution, throat slitting, and CO_2 stunning, which is presented as a "more humane" way to kill. It boggles my mind that some people studying the welfare of "food animals" can use the word *humane* to describe the unspeakable pain and suffering these sentient beings endure.

I don't want to go into detail about these horrors. However, if you want a glimpse of what this entails, read the 2023 *New York Times* essay entitled "Spy Cams Show What the Pork Industry Tries to Hide" by two-time Pulitzer Prize–award winner Nicholas Kristof. In sickening detail, Kristof tells it all and provides a link to the footage, which is quite graphic and upsetting. He writes, "The hog industry insists that the process is humane, so watch this video clip and decide for yourself."

Thankfully, this unrelenting cruelty is being increasingly noticed and leading to action. However, so far, the focus is primarily on animal welfare, or making the lives of the animals killed for food marginally better, rather than on animal well-being, or respecting the sentience and suffering of each individual animal. Too many welfare efforts are simply "humane washing." They are not aimed at truly improving the lives of animals but at helping us feel better about the cruelty that is served on our plates.

For instance, the movement toward raising "free-range" chickens and livestock is one way farms are trying to mitigate the worst abuses. That isn't wrong, but these animals are in no sense "free" and they are still killed in cruel ways. One of my colleagues says that the main question that motivates people in the poultry industry is, "How many chickens can you get in a cage with Vaseline and a shoehorn?"

More recently, in 2023, a new animal welfare scoring system was being developed to promote better pig farming practices. This would allow for better comparisons among pig farms about how the pigs are raised and killed. As one article said, "Making farming systems more sustainable, in the face of growing global demand for meat, is a major challenge for farmers and the government." Then it described how pork labeled "Woodland" would indicate farms that provide at least partial tree cover for the pigs, and the label "Organic" would indicate farms that provide outdoor access. In theory, these are tiny steps in the right direction. Another supposedly positive example is the use of warmer incubators for chickens, which can make male chickens more sociable, but how meaningful are these changes for the animals? Do they truly improve well-being, or are they little more than self-serving, feel-good gestures to soothe the consciences of researchers, producers, and consumers?

Lots of people defend current agricultural practices as humane. Some veterinarians have called pigs "gastronomic godsends" because they eat just about anything and all they have to do is produce meat. Margo Andrae, the chief executive for Australian Pork Limited, claims, "Everything we do is about minimising the stress on those animals, so they have a good life, and the CO_2 stunning is one of those ways of minimising harm. In terms of the product...ensuring that we minimise the stress is actually better for the products that we are privileged to serve to Australians."

Governments are also changing laws and regulations. In May 2023, the US Supreme Court supported California's Proposition 12, which requires the living space for sows to be increased from sixteen to eighteen square feet to twenty-four square feet per sow. Of course, pork producers strongly oppose such a tiny increase because actually making more space for the pigs they kill costs money. Again, increasing space for pigs is movement in the right direction, but this change is minimal and won't do much to improve the well-being of individual sows. This humane-washing move does not come close to the real solution, which is ending the practice all together.

To me, there is no way to make the current lives of food animals more humane, not without entirely revolutionizing farming practices or, ideally, not eating meat. I take issue with some of the proponents of animal welfare in farming, like Temple Grandin, an iconic welfarist and apologist for industrial agriculture, who has famously said, "I think using animals for food is an ethical thing to do, but we've got to do it right. We've got to give those animals a decent life and we've got to give them a painless death. We owe the animal respect." That sounds nice, but Grandin's vision is never realized in practice. Further, incremental efforts to improve welfare only obscure the depths of the problem, and millions of animals live horrific lives and die horrific deaths as people work to make their lives better rather than working to shut down these torture chambers.

To me, respecting animals means not torturing them and then killing them "kindly." Yes, some companies and governments are occasionally stepping in to make the lives of food animals "better," but "better" in no way means they have anything close to a marginally "good" life. Considering the unbelievable scale of the problem, these minor Band-Aid efforts are too few and far between, even if they are at least pointed in the right direction.

The Vegetarian/Vegan Solution

As we talked of freedom and justice one day for all, we sat down to steaks. I am eating misery, I thought, as I took the first bite. And spit it out.

— ALICE WALKER, "AM I BLUE?"

I decided to become a vegetarian and then vegan solely for ethical reasons. It was easy to do and I went "cold tofu" overnight. I didn't want to be part of the chain of inhumane treatment that slaughters sentience and legitimizes the factory farming of beings from cows to pigs to chickens to fish to lobsters. Many people are also surprised to learn of the horrors of the dairy industry. The dairy industry is anything but benign, and the veal industry would not exist without the dairy industry.

Many people are deeply conflicted about the food they eat. Who we eat, how animals are raised, how they are processed, where they come from, who profits, who's exploited: There is a wide range of complex issues to consider, and individuals must decide for themselves what is the most healthy for their own body and soul. Some people consider themselves vegetarians, although they aren't strict: They may still eat fish, or perhaps occasionally eat other meat (such as for a holiday meal), or restrict themselves to "animals without faces." For example, some people are willing to eat animals such as scallops or mussels because they believe they're not sentient and do not suffer. This may or may not be true, but it's where some people draw the line for themselves. Rather than vegetarians, these people are what Peter Singer and Jim Mason call "conscientious omnivores" in their book *The Way We Eat*.

I realize that, despite my wishes, ours isn't going to become a vegetarian or vegan world anytime soon, if ever. Yet I also don't understand why arguing for one is considered "radical." Is it radical to recognize that animals have emotions and to respect and honor their passions? I believe a vegetarian or vegan world would be a more compassionate world. While it is very difficult to track accurately how many more people are becoming vegetarians or vegans, both types of meal plans are gaining in popularity. It's also important to stress that a "vegan ethic" is a practical one that informs the fair and just choices people constantly make in many different areas. As Emilia Leese and Eva Charalambides point out in *Think Like a Vegan: What Everyone Can Learn from Vegan Ethics*, vegan ethics aren't only about food. Rather, they are a holistic approach that informs the choices we make in numerous situations, including politics, law, meal plans, friendships, and love in which fairness and nonhuman animals are involved.

For many people, eating meat is emotionally and ethically challenging. Many people feel a cognitive dissonance associated with eating animals, and these debates have moved from academic ivory towers into the homes of people worldwide. In his outstanding book *The Meat Paradox: Eating, Empathy, and the Future of Meat*, Rob Percival, an expert in the politics of meat, raises the central paradox: Given that we dislike causing harm to animals, why do we consume so much meat, and mostly from industrial farming systems, which cause so much harm? Percival analyzes the psychological and emotional forces shaping our diets, and the coping strategies that we commonly employ to navigate the tension. He asks, "At what point in our evolution did we begin to sympathise with the animals we consumed? After all, is this not peculiar? The lion does not mourn the gazelle. The wolf offers no reparations to the caribou, at least none we can recognize. But we might. What happened to us?"

I try to capture the same thing when I tell people we should think of a bacon, lettuce, and tomato sandwich as a Babe, lettuce, and tomato sandwich. When I say this to people, particularly youngsters, they remember that the bacon had been a sentient being no different than the movie star pig Babe. We recognize sentience and cherish it, but we often disassociate that knowledge from our food.

The bottom line is that there aren't many straightforward answers. Percival aptly notes that his research shows "our relationship with meat to be more complex, contradictory and consequential than has previously been told." I hope, as we continue to highlight the tensions that people feel, more and more people will cut back and eventually eliminate meat from their diets.

At Racetracks and Rodeos: That's Entertainment

All kinds of animals are used for amusement, performances, sports, and entertainment, which includes circuses, rodeos, pigeon shoots, bullfighting, rattlesnake roundups, fishing, polo, wildlife-killing contests, the Iditarod, and much more. As in the food industry, there are too many examples to cover in this book, so I'll look briefly at two: horse racing and rodeos.

Consider the Kentucky Derby. Horses who run at this huge accident- and death-ridden pageant are physically and emotionally abused and greatly suffer. They have no rights or say about what they're forced to do. Just before the

2023 Kentucky Derby, seven horses were already dead. Shortly thereafter, the total fatalities increased to twelve, and people were talking more about dead horses than who won the race. At the Santa Anita Racetrack, forty-two horses died in 2019. Then, after so-called safety improvements were made, twenty died in 2020, twenty-one died in 2021, and twelve died in 2022. The injuries and deaths only tell part of the story, however. Racehorses lead horrific lives and get nothing for their life-threatening efforts, since they certainly don't care about winning or losing.

Rodeos also are abusive for the nonhuman animals who are forced to entertain their audience. Calves are routinely and brutally abused when they're "roped" in rodeos. While some people say that the distress experienced by calves calls for better welfare, I favor banning roping and rodeo altogether. Nothing would be lost, as there is no benefit for rodeo animals. A few years ago I had a very good chat with a rodeo fan. I asked them, "What's in it for the animals?" They broke eye contact and said, "Absolutely nothing. When the human performers are through with them, they simply go on to find other animals to train, restrain, and use until they're no longer useful. It's the same old cycle over and over again." There also are rodeo-style chicken roping contests for entertainment at bars, in which the poor birds are abused as if they're merely unfeeling objects. Here is one description of this heinous "sport": "It's exactly like calf roping.... They run a chicken out of a little pen and they rope it.... Someone throws a cord around the neck and someone [else] tries to get one of the feet." The two participants then pull the chicken in opposite directions.

It's Not Happening in the Zoo: Cages and Conservation

But whatever thrill is to be derived from staring at a captive tiger is quickly dispelled by the animal's predicament. Awe gives way to abashment and then to a nearly inexpressible loneliness over being the only beast that does this to another. As such, any zoo, in whatever form, becomes not a demonstration of our prowess so much as a pathetically confused and protracted apology made to a series of wholly diminished and uninterested subjects.

— CHARLES SIEBERT, "THE DARK SIDE OF ZOOTOPIA"

People are entertained by common and exotic animals at zoos, but is this reason enough to put animals through the tribulations of being trapped and taken from their natural environments, separated from their families, housed in cages, moved around as if they're merely breeding machines, and put on public display seven days a week, including holidays, with little to no privacy? No, which is why zoos operate with two express purposes: One is to educate the public about animal behavior and conservation, and the other focus is on conservation and the preservation of species.

Education and conservation are wonderful goals, but they rest on two shaky premises. One is that zoos can actually succeed at them, and the other is that zoos can adequately care for their charges. As for their goals, there is insufficient evidence to know the extent to which zoos actually educate visitors or if zoos play any significant role in species protection. Even the Association of Zoos and Aquariums (AZA), which oversees US zoos and grants accreditation if zoos meet certain standards, has acknowledged in its own executive summary that "little to no systematic research has been conducted on the impact of visits to zoos and aquariums on visitor conservation knowledge, awareness, affect, or behavior." This was true when I first wrote this book and it remains true today, despite claims otherwise. Dan Ashe, the current president and CEO of the AZA has said, "It has never been more evident that conservation of wildlife and wild places depends on winning the hearts and minds of people. People visit zoos and aquariums, so this is increasingly where we are going to protect wildlife."

Meanwhile, AZA's conservation and management program, called the Species Survival Plan (SSP), attempts to ensure the survival of certain wildlife species using managed breeding programs, conducting basic and applied research, training wildlife and zoo professionals, and reintroducing captive-bred wildlife to their proper habitat. These things certainly sound good, but there is widespread skepticism about their application. Terry Maple, former director of Zoo Atlanta, has been quoted as saying, "Any zoo that sits around and tells you that the strength of zoos is the SSP is blowing smoke."

Some zoos do have strong interests in conservation, but data do not support the idea that what people learn in zoos does much to help wild animals, and many people worry that focusing on the role that zoos might play in conservation actually distracts from what really needs to be done for wild animals. The bottom line centers on whether people who visit zoos give money to different animal causes or go into a career that focuses on helping animals. Surely, some do; people can be inspired to donate or even become a biologist because they went to zoos when they were young. But the vast majority of visitors do neither. Research published in the journal *Conservation Biology* shows that children are not well educated when they visit zoos. Although more people have shown an interest in conservation and biodiversity during the past decade, this isn't related to zoos, and there has been little to no increase in people doing something to save nature — interest isn't spilling over into action. Many zoos engage in "conservation washing."

So do the data justify keeping animals in captivity? No. Do the data justify the common practice in zoos of killing so-called "surplus animals" who don't fit into a zoo's breeding program, which I dub *zoothanasia*? No. Does the slight chance that a zoo might actually "reintroduce" some of their residents to the wild justify keeping an entire zoo's worth of animals in cages indefinitely? No. The truth is, reintroduction projects are very rare, since there is evidence that animals bred in captivity differ from their counterparts living freely in nature. Captivity results in physical and behavioral changes, such as shorter wingspans in birds, changes in skull bone shapes, and weaker bites in lions and other carnivores. These changes may compromise their ability to survive in the wild.

So if zoos don't really educate and aren't important for species survival, can they at least be trusted to nurture their animals? Unfortunately, the answer is often no. Research shows large mammals suffer brain damage in zoos. And too often we're intentionally misled about zoo conditions and the welfare of the animals who are dependent on our care. We must demand more, for far too many animals suffer every day in zoos. The results of an opinion poll taken in the UK and commissioned by Born Free showed that more than 75 percent of respondents voted to end the keeping of large animals such as elephants, lions, tigers, giraffes, and rhinos in zoos, wildlife parks, and other captive facilities. And in February 2023, Puerto Rico announced it will close their only zoo because of complaints about mistreatment of its residents.

The Elephant in the Room: Suffering in Plain Sight

Elephants continue to die in captivity, and the death of an elephant is a significant event for any zoo. It is always noticed by the public. In June 2006, an elephant named Gita died in the Los Angeles Zoo, and *Time* magazine immediately responded with an essay entitled "Are Zoos Killing Elephants?" It's a fair question. Elephants are emotional, very social, and like to roam, and by definition, zoos are antithetical to these needs.

For instance, in spring 2001, Asian elephants were regularly moved in and out of the Denver Zoo as if they were couches being moved from room to room. The Rocky Mountain Animal Defense and I got involved because of the lack of concern demonstrated by both AZA and the Denver Zoo. To wit, Dolly, a thirty-two-year-old female, was removed from her friends, Mimi, forty-two, and Candy, forty-nine, and sent to Missouri on her "honeymoon," as the zoo called it, to breed. A few months later, Hope, a mature female, and Amigo, a two-and-a-half-year-old male (who had been taken from his mother), came to the Denver Zoo, where they lived next door to Mimi and Candy. In the following months, Mimi got increasingly agitated; in June 2001, she pushed Candy over, and Candy had to be euthanized. Two days after Candy died, and a day after she was autopsied within smelling distance of the other elephants, Hope escaped from her keepers and rampaged through the zoo. Luckily, no one was seriously injured. Hope was then transferred out of the zoo, and a new elephant, Rosie, was brought in.

Elephants live in matriarchal groups in which social relationships are enduring and deep. Their memory is legendary. Elephants form lifelong relationships and grieve when bonds are broken because of separation or death. When elephants move in and out of groups, there can be severe disruption of the social order and individuals can get very upset. This is what happened at the Denver Zoo. Mimi reacted to Dolly's leaving, and Hope reacted to Candy's death and autopsy. Were these happy elephants? Absolutely not. Yet the way they were treated is standard operating procedure for zoos, and it led to suffering and death.

Broken Families, Lost Friends

What is true for elephants is equally true for most other animals. Zoos can't simply tend to an animal's physical needs and call it a day. Animals also have social and emotional needs, and just as in humans, disrupting or ignoring them leads to negative consequences. Animals have families and friends, and they know when they are missing.

I witnessed this myself when, over the course of eight years, I and my students studied coyotes living in Grand Teton National Park in Wyoming. One female we called "Mom," since she was mother and wife, began leaving her family for short forays. She'd take off and disappear for a few hours and then return to the pack as if nothing had happened. We wondered if her family missed her when she left, so we observed what happened. Eventually, Mom's forays lasted longer and longer, often a day or two, and before she'd leave, some pack members would look at her curiously — they'd cock their heads to the side and squint and furrow their brows, as if to ask, *Where are you going now?* Some of her children would even follow her for a while. Then when Mom returned, they would greet her effusively by whining loudly, licking her muzzle, wagging their tails like windmills, and rolling over in front of her in glee. Clearly, her kids and mate missed Mom when she was gone.

One day Mom left the pack and never returned. The pack waited impatiently for days and days. Some coyotes paced nervously, expectantly, and others went off on short trips only to return empty-handed. They would travel in the direction she'd gone, sniff in places she might have visited, and howl as if calling for her. For more than a week, some spark seemed to be gone. Her family missed her. We believe the coyotes would have cried if they could. Their behavior told a story of deep, complicated feelings.

After a while, life returned to usual, almost. A new and unfamiliar female joined the pack, formed a partnership with the dominant male, and gave birth to ten coyote pups over two years. She became mother and wife. But every now and again it seemed that some of the pack members still missed their mom. The coyotes would sit up, look around, raise their noses to the wind, go off to look for her on short trips in the direction in which Mom had disappeared, and return weary and alone.

It was three or four months before these searches ended. Perhaps, despite their feelings, enough was enough. The life of the pack had to continue. For coyotes in the wild, this recovery and healing from loss is possible. For coyotes held in the artificial confines of a zoo, it is infinitely more difficult. Zoos ship animals from one zoo to another for economic reasons, or simply to redecorate their exhibits in order to keep the public interested, but the vast majority of animals form close social and familial bonds, and all that moving around can have a devastating impact on an individual's well-being.

Emotional Enrichment: Tending to the Heart

As I've said, many, even most, zoo employees care deeply about the animals in their care; most often, animals suffer because their needs don't match the needs of the zoo business, not because their everyday human caretakers are cruel. In fact, there are always folks who recognize the emotional needs of these sometimes desperate animals and find ways to soothe and help them. Emotional enrichment can simply take the form of plain human contact, which doesn't have to be direct or physical. Simple gestures of acknowledgment, attention, and caring are enough to reassure the animals and foster change, and it works for animals of different personalities and various emotional needs.

Then there is a zoo in the Netherlands that once thoughtfully tended to the emotional needs of captive orangutans in Indonesia, where these apes are kept in small cages because loggers have destroyed their homes. Webcams and monitors were set up in both places so that orangutans in Indonesia could meet other orangutans in the Netherlands. And it seemed to work based on the gestures and facial expressions that the orangutans directed toward the screens.

Here are some positive elephant stories. In 2006, Suma, a forty-five-year-old elephant in the Zagreb Zoo in Croatia, was extremely upset after her partner, Patna, died of cancer. Soon after, Suma's keepers discovered that she took comfort in Mozart's music — for she relaxed and "leaned against the fence, closed her eyes, and listened without moving the entire concert." The music helped Suma cope with her grief, and zoo authorities bought a stereo and provided music therapy for her. Music can soothe and calm down many animals. My sister, Margie, plays harp for cats and dogs in a shelter in Middlebury, Vermont, to relieve their stress, and it works well.

Ultimately, my hope is that one day all elephants will be removed from zoos and kept in more hospitable sanctuaries, the same way Ruby was moved from the Los Angeles Zoo to a sanctuary to live out the rest of her life in comfort. Since 1991, more than thirty US zoos have closed their elephant exhibits, including the Philadelphia Zoo, the San Francisco Zoo, and the Detroit Zoo. Other zoos have plans to close their exhibits, but change can be slow. In 2006, the *New York Times* reported that the Bronx Zoo "had announced that upon the death of the zoo's three current elephant inhabitants, Patty, Maxine, and Happy, it would phase out its elephant exhibit on social-behavioral grounds — an acknowledgment of a new awareness of the elephant's very particular sensibility and needs." However, elephants can live a long time, and as of spring 2023, Patty and Happy were still alive and still living in the zoo. So that year, Brooklyn council member Shahana Hanif, working with the Nonhuman Rights Project (NhRP), proposed a bill that, if passed, would be the nation's first ban on keeping elephants in captivity. In 2022, the NhRP also brought a landmark case before the New York Supreme Court requesting that Happy be granted legal personhood as a "rights holder." They lost the decision, but are appealing the ruling, and if it succeeds, it might pave the way to grant the same legal status to other social mammals, like gorillas and dolphins.

A Prescription for Zoos

Zoos likely are here to stay in the short term, but I favor phasing them out and moving their residents to more favorable environs. The Association of

Zoos and Aquariums should also enforce higher standards for accreditation and reaccreditation. Money saved from not holding animals in captivity could go toward protecting animals in their natural habitats, preserving critical habitat, and humane education. But while zoos remain, we must satisfy the physical and emotional needs of all their residents by rewilding zoos and offering animals the best lives possible in as natural an environment as possible. We must make their needs and emotional lives the top priority — they are the reasons zoos exist.

In a 2018 article called "A Postzoo Future: Why Welfare Fails Animals in Zoos," Jessica Pierce and I offered six suggestions for reforms to change the ethical landscape and bring zoos into a more ethical future. These prescriptions would also apply to captive animals in aquariums, which should ultimately be replaced by sea sanctuaries. These ethical changes include the following: (1) Shut down bad zoos; (2) stop exhibiting animals who cannot and never will thrive in captivity; (3) stop killing healthy animals; (4) stop captive breeding; (5) stop moving animals around from one zoo to another; and (6) use the science of animal cognition and emotion on behalf of animals.

Freedom is the key variable, and the root of the problem is captivity itself. As Dale Jamieson aptly puts it, "Zoos deny agency" — the freedom for animals to make important choices. Sure, some animals can make some choices, but they're still incarcerated and their choices are very limited. This relates to the concept of consent: In their own ways, all animals in our care should be allowed the ability to agree or disagree with how we choose to treat them. Consent underlies the development and maintenance of friendly and happy relationships among people, and it should be so within animal-human relationships. This includes every venue and activity in which humans and other animals interact, such as companion animals. Pets should be allowed agency. We should ask for their consent when we want to play with them, take them to the veterinarian, hug them, and so on. For all beings, agency and consent have positive effects on our emotional lives and support our needs for safety, security, and trust, and they enhance social bonds.

Let's briefly consider these six suggested reforms for zoos.

REFORM 1: SHUT DOWN BAD ZOOS

The majority of zoos are horrific. Bernard Harrison, past director of the Singapore Zoo, was asked in a 2017 interview if he agreed that most zoos are not fit places for animals. "I couldn't agree...more," he said. "If I could, I would shut down 90 percent of the ten thousand zoos in the world. The world is full of horrible stink holes that call themselves zoos." Jenny Gray, chief executive officer of Zoos Victoria and author of the book *Zoo Ethics: The Challenges of Compassionate Conservation*, wrote: "Unfortunately, the bulk of zoos in existence today still fall short of meeting the requirements of ethical operations. At best, 3 percent of zoos are striving to meet ethical standards, with perhaps only a handful meeting all the requirements." If these estimates are accurate, that means at least 90 percent of all zoos in the world should be closed immediately, since they do not meet the current standards for animal welfare.

REFORM 2: STOP EXHIBITING ANIMALS WHO CANNOT THRIVE IN CAPTIVITY

For some species, the deprivations of captivity are too profound, so another way zoos can make meaningful moral progress is to stop keeping these animals. This includes a well-known list of popular species who simply do not and cannot thrive in the zoo setting, such as elephants, bears, wolves, whales, dolphins, chimpanzees, orangutans, lions, and tigers, just to name a few.

REFORM 3: STOP KILLING HEALTHY ANIMALS

Many people don't know that zoos routinely kill healthy animals in their care, and when people find out about this practice, they are appropriately upset and sickened. Typically, zoos kill what they consider "surplus animals," those they don't need for breeding programs (and so they don't need their genetic material) and who have simply become too expensive to keep. Zoos call this euthanasia, but this is not a compassionate death for animals who are terminally ill or in unbearable pain; it's what I call *zoothanasia*, the slaughter of healthy animals who did not ask to be caged in the first place.

This is not a small problem, but it's hard to know the exact extent, since zoos don't like to publicize the practice. A 2014 *BBC News* story said that the European Association of Zoos and Aquaria "does not publish these records or advertise the number of healthy animals that have been culled, but executive director Dr. Lesley Dickie estimates that somewhere between three thousand and five thousand animals are 'management-euthanized' in European zoos in any given year."

Nevertheless, cases of zoothanasia garner attention: In 2014, Marius, a healthy eighteen-month-old giraffe was killed at the Copenhagen Zoo, which four weeks later killed four lions. In 2018, it was discovered that a zoo in

Sweden had killed nine healthy lion cubs since 2012. In December 2022, three chimpanzees were shot dead soon after they escaped from their cages in the Furuviksparken zoo in Sweden, and a fourth shortly thereafter. The zoo was closed at the time and said the chimpanzees were killed because the zoo didn't have enough anesthetic on hand and out of concerns for human safety. And in April 2023, the Dundee Zoo killed an entire healthy wolf pack after they had to put their leader to sleep after complications from surgery. Zoo workers made the decision to kill the surviving pack because the wolves displayed "unusually anxious and abnormal behavior" after their leader was killed, but this is a normal reaction. Typically, packs break up after a leader dies, though this option was obviously unavailable. In this way, zoos create the conditions for problems that they wrongly "solve" by killing animals.

REFORM 4: STOP CAPTIVE BREEDING PROGRAMS

One way to end the practice of killing healthy animals is to end captive breeding programs. Instead of fretting about the ethics of what to do with healthy animals, zoos should simply stop breeding animals until they have too many to afford to adequately care for them. Making the "hard decision" to zoothanize healthy animals is not forced upon zoos. Zoos themselves create the problem, and if they alter their practices, this moral conundrum would not arise.

REFORM 5: STOP MOVING ANIMALS AROUND AMONG ZOOS

Zoo directors provide numerous explanations for why animals must be lent out like library books among various institutions. Elephants and giant pandas are among the animals shipped around on loan with little consideration for the emotional impact on the animals themselves, such as what happened when the elephants Dolly, Mimi, and Candy were moved among zoos.

One reason animals are moved is to support breeding programs, which zoos claim is both good for business and essential for conservation. For captive pandas, this amounts to a breeding boondoggle that is difficult to fathom, as some zoos use forced methods for gathering semen from males and impregnating females that are barbaric. Baby animals are especially useful for zoos to make money. British zoologist Lucy Cooke calls baby pandas "cuteness crack." But children, too, can be unnecessarily killed when, for example, their mother dies. This happened in early 2023 at the Basel Zoo, when a four-day-old baby orangutan was killed after her mother died.

Zoos also make money selling animals (or their semen) to other zoos. Yet the justifications don't really matter. Whether animals are brought in for an exhibit, or are brought in for their DNA, or are moved because one zoo has a surplus and another has a deficit of a given type of animal, and each needs to balance their exhibits, every reason fails on moral grounds. There is no good animal welfare–based reason for trading and moving individuals, and if zoos are going to become moral institutions, animal-based reasons must drive practice. If animals in zoos really are "residents," which is the current term of art within zoo management circles, then we should consider their home zoo as just that — their home.

REFORM 6: USE THE SCIENCE OF ANIMAL COGNITION AND EMOTION ON BEHALF OF ANIMALS

This last reform is about bridging the "knowledge translation gap." It's long past time that zoos, and society in general, put into practice what we've learned from the scientific study of the emotional and cognitive lives of animals. As research continues to confirm what we already know, practical consequences are lacking. This gap is nowhere more evident than in the realm of zoos.

All in all, even so-called "good" zoos have a lot of work to do to give their residents better lives. Even then, a "better life" in a zoo is not necessarily a "good life."

In the Wild: Civilization and Its Discontents

We are part of a generation of shame, where future generations will look back on our treatment of animals and be appalled.
— JILL ROBINSON, FOUNDER OF ANIMALS ASIA

Domesticated humans have a long and troubled history with the wilderness that exists at the edges of their civilization. We try to tame it and keep it at bay, but we almost never just leave it alone. Perhaps we can't, and it's not really our fault. Wilderness is considered "wild" for a reason: It grows where it likes and has its own rules, and it rarely respects the artificial lines — the fences and park borders — we draw to contain it.

Though wild animals would seem to fall outside the responsibility of humans, we in fact need to care for them, too — precisely because we already try to "manage" wilderness. We do this primarily for our own benefit, although some people pretend that they're putting the lives of nonhumans first. As our cities and suburbs grow, as we claim more and more acreage for grazing and farmland, wilderness shrinks, and this habitat loss can be deadly. Animals who are accustomed to ranging for hundreds of miles struggle with these limitations, and it forces them to encroach on our human civilization. When they do, we don't like it: Deer ruin our gardens and predators kill our cows, sheep, and other animals we pejoratively label as "livestock." So we kill certain wildlife to protect what we've built. In most of the American West, native predators, such as grizzly bears and wolves, have been eradicated. It's a vicious cycle.

One way to mitigate this would be to pay more attention to how and where we build and live. As much as possible we need to account for the natural habitats and habits of wild animals to ease these conflicts, which are inconvenient and expensive to us and deadly for them. A workable strategy is to preserve wildlife corridors, such as those proposed by the Yellowstone to Yukon (Y2Y) conservation initiative. Y2Y is establishing a system of corridors that enable grizzlies and other animals to move freely in this mountain ecosystem.

Numerous wild animals are killed by federal agencies. In 2022, Wildlife Services ruthlessly killed more than 1.75 million animals, around 200 per hour, including 404,538 native animals. They also slayed almost 3,000 animals by accident in 2021. In Washington State, the hired guns of Wildlife Services kill more than a hundred thousand animals each year. Experts dispute the necessity of all this killing, and typically, Wildlife Services refuses to comment publicly on specific incidents, except to defend the agency's methods as "biologically sound, environmentally safe, and socially acceptable." What an egregious lie.

Killing is usually done through traps, leg and neck snares, poison, shooting animals from aircraft, and pulling them out of dens. While the total number of animals killed has declined over the past few years, it should be stopped, since mounting evidence shows that such killing usually doesn't solve whatever problem they've identified. This war on wildlife is utterly despicable, and what makes it worse is that it's done in the name of coexistence, as a way to solve human-animal conflicts.

In her well-researched and forward-looking book *Beyond the North American Model of Wildlife Conservation: From Lethal to Compassionate Conservation*, Anja Heister clearly shows why the violent North American model of wildlife management fails other animals and us and violates even a minimalist code of ethics. She offers an informed and passionate critique of the dark side of government-orchestrated exploitation and destruction of wildlife. She argues that we need to replace power and violence with nonlethal coexistence and that we need a new psychological framework that shifts from anthropocentrism to concern, empathy, and respect for the lives of individual nonhumans.

In my home state of Colorado, deer and elk suffer from chronic wasting disease, and in 2001 the state worried that continued spread of the disease would harm its hunting industry. Therefore, the Colorado Division of Wildlife (now called Colorado Parks & Wildlife) began a program of deer and elk killing with the intention that — follow the logic here — more animals would survive so hunters could kill them. However, by 2006, local newspapers reported, "Officials acknowledge that killing deer and elk to contain the spread of chronic wasting disease hasn't worked." Unfortunately, they reached this conclusion only after twenty-three hundred animals had been killed by the program.

Using the same logic, others argue that wolves are predators who kill too many elk, so the wolves should be killed to ensure that hunters have enough animals to shoot at. This myth has a long history, and even in 2023, it is still being used by those who oppose bringing wolves back to Colorado. If anything, prey animals belong to their predators, along with whom they've co-evolved; wild animals aren't "our animals" to kill. Further, studies have shown that wolves don't significantly decrease populations of elk, deer, and other ungulates. Wolves typically kill weak and sick individuals, whereas hunters often seek out trophy animals who are breeding stock for future generations. Wolves have also been killed to protect livestock, but short of killing every single wolf (which nearly happened by the end of the nineteenth century), this effort is also ineffective. The same is true for coyotes; killing them to control their predatory habits doesn't do much because they really don't kill as much as many claim. Killing wildlife for fun also doesn't work to reduce predation. While "killing for fun" wildlife contests are still legal in most of the United States, one positive move occurred in 2020, when Colorado joined Arizona, California, Maryland, Massachusetts, New Mexico, Vermont, and Washington in ending these heinous bloodbaths.

Whatever short-term benefits killing wild animals provides, these don't work for the long term. Killing is easy, but it's not a lasting solution. Wild animals abide by their evolutionary behaviors, and they adapt to the encroachments of our civilization. It is only by our definitions that they become "pests," "trash," or "nuisances." As Bethany Brookshire shows in her 2022 book *Pests*, we create animal villains by labeling them pests, but this says more about us than about them. The term *pest* is convenient because it allows us to get rid of unwanted animals without feeling guilty. It is incumbent on us to curb ourselves, our expectations, and our "boundaries" to reduce conflicts with wild animals. Peaceful coexistence doesn't mean eliminating the other side.

Honest Talk and Rewilding Language

I constantly think about the words we use to refer to nonhuman animals. Our language both embodies and influences how we view animals, our attitudes, and how we treat them. Our language can feed into a speciesist mindset, one that diminishes and objectifies other animals, and it can reflect respect for the minds, emotions, and inherent subjectivity of animals, including individuals. One ethical issue we all face, but one that we often aren't aware of, is making choices about the language we use. We must pay attention to our words and decide if they reflect our values. If certain words don't, then we should replace them with ones that do, as well as calling attention to when others use language that objectifies animals and sanitizes harm.

Sometimes, when we experience cognitive dissonance — such as when our actions toward animals don't reflect our beliefs or feelings toward them — we psychologically distance ourselves from other animals by using euphemistic, indirect language. Wildlife and conservation scientist Dr. Gosia Bryja aptly notes that when people sanitize their language to justify killing or harming other animals, or to make it sound palatable or justified, this is a way of "silencing nature" — like applying a knife to an animal's vocal cords. Bryja's 2023 essay "Language and the Knife: Silencing Nature" should be required reading for everyone, but especially those who work and interact with animals directly.

In this book's introduction, in "A Note about Terminology" (page 12), I acknowledge some of the choices I make when talking about nonhuman animals. I use *who* and *they* rather than *it*, *that*, or *which*. That is, I use the same subjective pronouns we use for people, not the pronouns we use for things and objects. I've also raised the issue of language in other places, such as by noting that, when people choose to eat meat, it's a matter of *who's* for dinner, not *what's* for dinner. This calls attention to the fact that the animals who wind up on our dinner plates were once alive and sentient. They are not objects, and our words should not objectify them. I know because people have told me that this small change in language has resulted in a number of people deciding to change to vegetarian or vegan diets.

One way we can make the world a kinder place is to "rewild" our language so that it honors and respects the sentience of animals. Language, thought, and action are well-connected, and changes in language could, and should, change how we interact with other animals. The solution to mislabeling animals, and what we do to them, is to watch what we say, and if enough people change how they speak, such "rewilding" will seep into scientific literature, mass media, and everyday conversations. When people use a misnomer, gently correct them and explain why. This can lead to fruitful discussions about animal-human relationships. Nothing is lost and a lot can be gained that helps foster genuine coexistence.

Gosia Bryja says this image is composed of "falsifying words," or terms for nonhuman animals that diminish them or deny who they are — sentient beings. *(Image used by permission of Gosia Bryja.)*

Sanitizing the Language of Killing

Perhaps the most important issue is the language we use to justify and sanitize killing. As with scientists and zoos, people rarely mind admitting to the sentience and subjectivity of animals when those animals are clearly happy — especially when that happiness is directed at us. We want to feel loved by our dogs, cats, and other animal friends. But people are often uncomfortable admitting when animals are unhappy and in pain — especially when we are causing their suffering.

Legally, only humans can be murdered, but that is speciesism. What is it other than murder when zoos "euthanize" healthy animals because they've run out of room and money to keep them? Parents aren't allowed to "euthanize" babies if they decide they already have too many.

Instead of *killing*, people use a wide range of euphemisms: *sacrificing, culling, removing, harvesting* (we harvest corn, not animals), *bagging, taking, dispatching, disposing, gathering, eliminating*. In the case of wolves being killed in Washington State, I came across the phrase *authorized removal*. No doubt, at least some of the people authorized to remove the wolves might balk at the term *sanctioned murder*.

As for the animals who are killed, they are often villainized and transformed by language into something that needs to be gotten rid of. Animals are called *pests, vagrants, livestock, game, surplus, prey, nuisance species, invasive species*, and more.

The list goes on. These words don't reflect who these animals truly are or what people do to them. For instance, take the phrase "ventilation shutdown." This is a term for a method of mass animal "depopulation." In theory, this is only meant to be used to stop the rapid spread of disease in chickens, pigs, turkeys, and so on. During this disgusting, heinous process, a barn is closed, the airflow is turned off, and the heat is raised until the animals die of either hyperthermia (overheating) or suffocation. Death can take hours, and those who live longest must witness the deaths of everyone else. Dr. Karen Davis, president and founder of United Poultry Concerns, argues that, before they die, chickens suffer from "traumatic empathy" because of having to witness other chickens being killed. *Ventilation shutdown* is certainly a more neutral, nonjudgmental term than *mass murder*, though it's hard to imagine that the people who engage in this don't recognize what it is.

When it comes to hunting, two terms some people use are *mutualism* and *ethical hunting*. These phrases are used to present hunting as both moral and part of the natural order, but they are really just ways for people to feel good about killing. Since wild animals hunt and consume one another, and this symbiosis is part of the circle of life, some people consider hunting part of "mutualism." But this is a misuse and perversion of the term, since there is nothing mutualistic about modern hunting, at least from a wild animal's point of view.

The phrase *ethical hunting* is often used by people who don't engage in trophy hunting — that is, killing for fun — and who try to kill animals as quickly and painlessly as possible. But "ethical hunting" is simply another good example of humane washing. It presents unnecessary killing as if it were a noble act. What this means is that, for example, when an "ethical hunter" shoots an animal, the animal dies instantly, and then they consume the animal, rather than leaving them or turning them into a wall hanging. If the person actually needed the animal for food — for literal survival, not just because they like the taste of "game," or wild animals — then that becomes a different sort of ethical question. But most hunters don't need to kill to live, and so this becomes simply a way to give themselves guilt-free permission to take another life.

Similarly, some hunters abide by the concept of "fair chase." Originally coined by the Boone and Crockett Club, "fair chase" refers to "the ethical, sportsmanlike, and lawful pursuit and taking of any free-ranging wild game animal in a manner that does not give the hunter an improper or unfair advantage over the game animals." Some of their fundamentals of hunter ethics include: "Attain and maintain the skills necessary to make the kill as certain and quick as possible; behave in a way that will bring no dishonor to either the hunter, the hunted, or the environment; and recognize that these tenets are intended to enhance the hunter's experience of the relationship between predator and prey, which is one of the most fundamental relationships of humans and their environment." We're also told, "Fair chase rules make sure hunters have no unfair advantage over wild game by balancing the skills and equipment of the hunter with the abilities of the animal to escape."

What's fair about a human with a high-powered rifle killing a deer, for example, who hasn't evolved antipredatory skills against such a weapon? Nothing at all. In the real world, prey animals have evolved a wide variety of different behaviors to counter being killed and eaten by nonhuman predators, but they can't legitimately defend themselves against human predators.

These examples just scratch the surface of these issues, which often boil down to an ethical paradox that is hard to face. Some animals we love and would do anything to protect and care for, and some animals we don't like for any number of reasons, and some people don't mind if those particular animals die. Sometimes, we feel both things about the same animal or species. So how do we reconcile the difficulties of coexistence? Not by using sanitized language that helps us avoid the truly hard choices coexistence entails.

Compassionate Conservation to the Rescue

Killing animals in the name of coexistence is an oxymoron — it doesn't work and nonlethal alternatives need to be developed and used globally. Since I wrote the first edition of this book, the ever-expanding field of compassionate conservation has emerged and rapidly grown worldwide, and it can help us along in truly working for and achieving coexistence with a wide range of animals, so-called "pests" included.

Compassionate conservation is based on the ethical position that actions taken to protect biodiversity should be guided by compassion for all sentient beings. Researchers in various disciplines — including biology, psychology, sociology, social work, economics, political science, law, and philosophy — have been working closely together and made substantial contributions to this rapidly growing international, transdisciplinary field, and there are numerous success stories. The four guiding principles of compassionate conservation are: *first do no harm*, *individuals matter*, *value all wildlife*, and *peaceful coexistence*.

Valuing individuals, not just species or populations, often called "collectives," means that killing certain individuals in order to protect or "save" other individuals or species isn't permissible; other protocols need to be developed and implemented. Kristy Ferraro and her colleagues correctly argue that, not only does every individual have moral worth, but individual variation within species is very important for implementing conservation strategies.

Critics of compassionate conservation argue that there are three core reasons harming animals is acceptable in conservation programs: The primary purpose of conservation is biodiversity protection; conservation is already compassionate to animals; and conservation should prioritize compassion to humans. I think these criticisms are wrong and don't reflect what compassionate conservation is all about.

Because of my work in this area, I was invited to become a member of the United Nations Harmony with Nature Knowledge Network, a program that recognizes the intrinsic value of nature and promotes a nonanthropocentric worldview. This program embodies the same values as compassionate conservation, which is not simply "welfarism gone wild," as I call it. It offers a completely new paradigm for thinking about ourselves in relation to wild animals and is reshaping conservation ethics as we travel through the Anthropocene.

Compassionate conservation recognizes that humans are stakeholders, too, when there are animal-human conflicts. In our challenging world, it is unrealistic to try to resolve conflicts without considering all of the stakeholders.

This does not mean that human interests are always more important than nonhuman interests. Rather, we need to develop solutions that honor the fact that we have intruded, often inhumanely and abusively, into the lives of animals. Examples of "First do no harm" include using nonlethal paintball guns to deter pesky baboons in Cape Peninsula, South Africa; using Maremma sheepdogs to protect livestock in various locales, since poisoning the predators changes the dynamics of ecosystems and reduces biodiversity; and using sheepdogs to protect little penguins from red foxes in Australia.

A wonderful project in India run by Wildlife SOS India stopped the horrific practice of using dancing bears for entertainment and money by showing the local people that they and the bears could benefit in other ways. Conservation biologists Kartick Satyanarayan and Geeta Seshamani noted:

> Wildlife SOS spearheaded a conservation success story in India by resolving the barbaric dancing bear practice in which sloth bear cubs were poached from the wild, brutally trained in inhumane ways, and spent their short tragic lives at the end of a four-foot rope dragged through towns and villages to earn for the indigent, nomadic community called the Kalandars. Wildlife SOS's initiative was to both rehabilitate the sloth bears held in captivity and the Kalandars themselves in alternative livelihoods. This in turn made a huge difference to the sloth bear population in the wild, helping in its conservation.... Attempts at resolution involve creating safe spaces for the animals (rehabilitation centers), teaching people behaviors which do not lead to confrontation with the animals in question (awareness and education), but most importantly to inculcate a feeling for the animals in question, emphasizing adjustment and acceptance of the existence of wildlife close to our human habitations.

Clearly, it is possible to work out conflicts that benefit all stakeholders.

Another project centering on human conflicts with leopards showed that a simple change of perception from "problem animals" to "problem location" resolved many human-leopard conflicts in human-dominated landscapes without having to kill the leopards. The research, carried out by wildlife scientists and park managers in Sanjay Gandhi National Park in Mumbai, India, also called into question the effectiveness of capture and removal or translocation to new areas. The report concluded:

Changing the focus from "problem animals," which has proved ineffective, to "problem locations" also changes the issues that need to be addressed and measures that need to be in place. It brings into consideration site-specific measures to reduce leopard visitations that may lead to negative interactions with people. This could include control of the domestic dog population, appropriate disposal of garbage and kitchen and medical waste, and better protection for livestock in safe corrals. It draws attention to providing better amenities for tribal and underprivileged people living in the area, including lighting, housing, sanitation (to reduce open defecation in forests, especially at night), and other public safety measures. Changing the focus from leopard to location also implies a change from reactive measures (such as capture after conflicts occur) to proactive efforts (such as making safer surroundings to preempt attacks) that reduce negative interactions while enabling people and leopards to share the landscape.

What's pertinent here is that emotions can be a source of moral understanding in conservation — they don't get in the way of finding effective solutions, and they are not "soft" and antiscience. Emotion and compassion should be embraced as core virtues of conservation. Our emotions and those of other animals matter, and we need to weave them together to work for all animals, who are trying hard to survive and thrive in an increasingly human world. While we're at it, we need to do some deep thinking about how we ever evolved to the point that we need to manage everything and everyone because of our self-centered anthropocentric misbehavior.

Reintroducing Wolves: Sentience, Emotions, and Conservation

Projects centering on reintroducing wolves to some of their original territory — as is going on currently in my home state of Colorado — offer excellent case studies to learn about and practice compassionate conservation. This was brought home to me when I read a January 2023 essay entitled "Human-Caused Mortality Triggers Pack Instability in Gray Wolves," which was written by a group of the world's leading wolf experts. This made me revisit some of the basic tenets of compassionate conservation and think more about what is involved in these extremely challenging projects. Examining a massive database consisting of more than 118 years of cumulative data from 5 national parks, all within the historical and current gray wolf range in the United States, the researchers analyzed 193 packs (with 2 to 37 wolves) over 864 pack years. These included 978 wolf mortalities from 1986 to 2021. What the researchers found is that when people kill wolves, packs becomes less stable and can disband and cease to exist.

Here is a brief summary of what the researchers learned.

- Human-caused mortality accounted for 36 percent of collared wolf mortalities and had a detrimental effect on both pack persistence and reproduction. The human-caused mortality of any wolf decreased the predicted odds of pack persistence to the end of the biological year by 27 percent and reproduction the following year by 22 percent.
- Packs with no reported human-caused mortalities persisted to the end of the biological year 91.6 percent of the time, while packs that experienced at least one reported human-caused mortality persisted 76.3 percent of the time.
- Packs with no reported human-caused mortalities reproduced the following year 79 percent of the time, whereas packs with at least one reported human-caused mortality reproduced the following year only 65.6 percent of the time.
- Larger packs were better able to recover from human-caused intrusions than smaller packs.

Of course, wolf packs come and go, and wolves leave their groups for a wide variety of reasons, but this shows how human intrusions and killing impact the surviving wolves — by increasing pack instability and reducing reproductive output. Clearly, more research is needed. For instance, we don't know how relocation programs — which remove wolves from packs in one location and move them to another location — influence the packs from which wolves are removed. It is extremely important to know a lot more about the effects on wolves of these types of human intrusions and experiments. This includes considering what wolves are thinking and feeling during translocation efforts, from the time they're pursued and trapped to when they're released into a new, unfamiliar place, where they're not always welcomed by us. The principles of compassionate conservation mean factoring in all stakeholders with every decision — including our feelings and those of the animals involved.

Philosopher Martha Nussbaum has argued that justice for animals means the freedom to do what comes naturally. I agree, and when animals can do what they've evolved to do in the presence of humans, it's a win-win for all — humans get to see these wild animals in action in the ways they would behave when humans aren't around.

It is incumbent upon researchers and nonresearchers alike to set an example for future generations who are striving for coexistence. We can look to a new generation of researchers who take seriously the value commitments of compassionate conservation. We cannot and must not continue to spill the blood of other species because it's convenient and supposedly gets the job done. As humane educator Zoe Weil aptly notes, the world becomes what we teach, and so we must focus attention on teaching children the values of compassion, generosity, nonviolence, and respect for others.

{AFTERWORD}

Compassion and Justice for All

Life supports life. Animals are the key.
Variety and abundance are the strengths. Animals could save us.
The paradox is, now, only we can save them.

— Keggie Carew, "Animals Can Save Us"

I want to end this book on a positive note. It's important to pay close attention to our successes on behalf of nonhuman animals, since it's far too easy to focus on the challenges and the harm that is still being done. Of course, animals need our help more than ever, but many good things are happening. Compassion fatigue and empathy fatigue can take their toll if we forget to appreciate progress and take care of ourselves.

Perhaps the most important thing is that caring for other animals is increasingly becoming a mainstream concern for a growing number of humans. It is no longer radical to recognize, respect, and want to protect the emotional lives of animals. Further, people increasingly see how we ourselves benefit by acting with more empathy, kindness, respect, and compassion for all the world's creatures. It can be true that other animals often suffer in silence, since they can't speak to us directly, but more and more people are learning how to read their behavior and speak for them.

While it is by no means a universal sentiment, I find that people are expressing more humility toward other animals and nature, perhaps as a consequence of the unmistakable damage our presence has caused. As David Attenborough rightly notes, "I think sometimes we need to take a step back and just remember we have no greater right to be here than any other animal." He also notes, "Sometimes we might remember that all other animals have every bit as much right to be here and to be unmolested as any human does."

All the research and studies of the last twenty years are helping us to recognize that caring for the well-being of nonhuman animals is the right thing to do. Caring crosses species borders. Good examples are the One Health and One Welfare initiatives, which are driven by the premise that caring for and working on behalf of other animals also helps us and is a way to care for ourselves. This caring allows people to rid themselves of the cognitive dissonance they experience over all the various types of harm that people cause to animals. Animal emotions are a matter of importance in their own right, but the very presence of animals — with their free-flowing emotions and empathy — is also critical to human well-being. As University of Denver's Dr. Sarah Bexell aptly puts it, "The One Health approach is a way of looking at the world that helps humans to see and acknowledge that humans, other species, and the natural environment are completely and perfectly interlinked. If we harm one of these three pillars, all three are harmed." Similarly, wolf expert John Vucetich and his colleagues note, "Caring for nonhumans, for their own sake, does not preclude caring for humans. Humans are more than capable of caring for many more than one kind of thing.... Nothing is inherently misanthropic about being nonanthropocentric."

If we continue to allow human interests to always trump the interests of other animals, we will never solve the numerous and complex problems we face. We need to learn as much as we can about the lives of wild animals. Our ethical obligations also require us to learn about the ways in which we influence animals' lives when we study them and what effects captivity has on them. As we learn more about how we influence other animals, we will be able to adopt proactive, rather than reactive, strategies. We should honor the rights of individuals to be free to live the lives they're meant to live — to be free to perform species-typical behaviors whether we like them or not. There's a certain wholeness when a wolf, coyote, eagle, robin, trout, or snake is allowed to be who they're supposed to be.

The fragility of the natural order requires that people work harmoniously so as not to destroy nature's wholeness, goodness, and generosity. The separation of "us" (humans) from "them" (other animals) engenders a false dichotomy. This results in a distancing that erodes, rather than enriches, the numerous possible relationships that can develop among all animal life. What befalls animals befalls us. A close relationship with nature is critical to our own well-being and spiritual growth. And independent of our own needs, we owe it to animals to show the utmost unwavering respect and concern for their well-being.

I find the word *solastalgia*, which was coined by Australian environmental philosopher Glenn Albrecht, to be appropriate here. It describes "the distress caused by the lived experience of the transformation of one's home and sense of belonging and is experienced through the feeling of desolation about its change." We experience solastalgia when we erode our relationships with other beings.

It's easy to become anthropocentric and forget that humans are fellow animals. Our species is different, but it's also the same. Theologian Stephen Scharper resolves this contradiction with his idea of an "anthro-harmonic" approach to the study of human-animal interrelationships. This view "acknowledges the importance of the human and makes the human fundamental but not focal." We are all in the world together. We and other animals are consummate companions, and we complete one another. Theologian Thomas Berry expresses this same idea a little differently. He says each and every individual is part of a "communion of subjects," in which our shared passions and sentience provide the foundation for a closely connected community. No one is an object or an other; we are all just us.

When we're unsure about how we influence the lives of other animals, we should give them the benefit of the doubt and err on the side of the animals. It's better to be safe than sorry. Many animals suffer in silence, and we don't even realize this until we look into their eyes. Then we know.

Personal Choices, Personal Rewilding

What should we do with what we know? What sorts of choices should we make? We *must* use what we know on behalf of all animals. I will admit, I get cranky and irritable now and again. I'm tired of reading studies and essays about animal behavior, animal cognition, animal emotions, and animal sentience that trumpet new discoveries and then end by saying something like, "We need to treat other animals better — with more respect, compassion, kindness, and dignity." Of course we do. These banal platitudes of pain don't do anything for me — why does the US Federal Animal Welfare Act still write off lab rats and mice as not being animals, and why, as of 2022, can more pigs be killed per hour in slaughterhouses than previously allowed? We need a breakthrough paradigm shift in how we treat other animals and a call for heartfelt action on their behalf, one that leads to changes in our laws, regulations, and animal-human interactions. That's what I hope this new edition will help achieve.

During this revision, I came across a lot of new material as well as some old material that I hadn't discovered before. All of it supports what I wrote in 2007 and have been writing about ever since. The only thing that's changed is that we've confirmed that more animals than we ever thought are emotional, sentient, and aware, and so the biodiversity of sentience has only continued to expand. Now it's time to use that knowledge to enforce existing welfare regulations and laws while also improving them, making them more stringent and with real consequences for violations, so people might think twice before choosing to harm and kill other animals. I hope this update inspires you to do something — anything, big or small — that improves the lives of nonhumans. Everyone can make a difference. That said, it's essential to work *for* animals and not *against* people. Through respectful coexistence, we can find a way to share our magnificent and fascinating planet.

As I was writing this book, I kept thinking about the adage "After all is said and done, a lot more is said than done." I hope what I've written inspires people to reduce the gap between saying and doing. Everything we have learned warrants this paradigm shift, and nothing we have learned calls for allowing continued or additional abuse. When we don't use what we know, we let animals down, and as a result we let ourselves down — what harms them harms us, and what benefits animals makes life better for all.

What this means is that we each must make our own decisions and choices and take responsibility for our own actions. Individual responsibility is critical. The problems that animals face, and that we face in caring for them, can be overwhelming. It's easy to become discouraged, it's easy to feel lost and powerless, and it's easy to put the blame on institutions and corporations, on "society," and fail to address our own behavior.

How do I decide what to do? Simply put, I try to make kind choices — choices that are fair and just. I try to increase compassion and reduce cruelty. I try to practice what I call the "thirteen Ps of rewilding" — being proactive, positive, persistent, patient, peaceful, practical, powerful, passionate, playful, present, principled, proud, and polite.

And I try to make it simple and fun. When I talk with people about what they can do to help animals and our wounded planet, I always stress how easy it is to do something positive — people don't have to found an organization or spend a lot of time. They can build small actions into their daily lives. I'm certainly far from perfect, but these goals motivate me daily. They guide me when I'm not sure what's right. When I challenge people, usually scientists, by asking, "Would you do it to your dog?" I'm really just trying to remind people to act with compassion. I'm trying to shock people into remembering the golden rule: Do unto others as you would have them do unto you. Who could possibly argue that all beings aren't deserving of the best lives possible?

To me, this includes personally rewilding our hearts from the inside out. We need to reconnect with other animals and with the magnificence of nature that forms their homes. We need to dissolve false boundaries to truly connect with both nature and ourselves, and one way to do this is to put aside our self-centered mindset. One of my favorite bumper stickers is "Nature bats last." We can try to outrun and outsmart nature, but in the end, she always wins. Will we allow ourselves to become one of the species that didn't make it? Or worse, will we continue to be the one species that threatens all others and who allows uncounted species and individuals to perish so we can live where and how we please? I hope not.

In practice, rewilding means walking through the world treating every living being like an equal — not the same, but as a being with an equal right to life. The golden rule applies to human animals, other animals, trees, plants, and even Earth itself. My colleague Jessica Pierce says that we need more "ruth," a feeling of tender compassion for the suffering of others. Ruth is the opposite of ruthless, or being cruel and lacking mercy. I agree. Kindness and compassion must always be first and foremost in our interactions with animals and every other being in this world. We need to remember that giving is a wonderful way of receiving.

Living up to this simple pledge isn't easy. Believe me, I know. To do so, we must overcome fear — fear of going against the grain, fear of coming out of the closet, fear of ridicule, fear of losing grant money or irritating colleagues, and fear of admitting what we've done or are doing to other sentient animals. Sometimes when we find it difficult to overcome our fears, a debilitating feeling of shame can paralyze us, but we must remember that every day brings new opportunities. No matter how small the gesture, anytime we act with compassion and do what we feel is right, despite the negative consequences (whether actual or imagined), we make a difference, and that difference matters.

In March 2006, I gave a lecture at the annual meeting of the Institutional Animal Care and Use Committee in Boston. I was received warmly and the discussion that followed my lecture was friendly, even though some in the audience were a bit skeptical of my unflinching stance that we know that certain animals feel pain and a wide spectrum of emotions. After my talk, a man came up to me who was responsible for enforcing the Animal Welfare Act at a major university. He admitted that he'd been ambivalent about some of the research that's permitted under the act, and after hearing my lecture, he was even more uncertain. He told me that he'd be stricter with enforcing the current legal standards, and he would work for more stringent regulations. I could tell from his eyes that he meant what he said, and he understood that the researchers under his watch would be less than enthusiastic about his decision. But he needed someone to confirm his intuition that research animals were suffering, that the Animal Welfare Act was not protecting them. I was touched and thanked him. Then he put his head down, mumbled, "Thank you," and walked off.

One notion I've been writing a lot about lately is "rewilding education." It meshes well with what humane educator Zoe Weil has said — the world becomes what we teach. One aspect of that is getting youngsters off their butts and out into nature. Also, we can encourage schools and parents to include humane education, so we raise children who both understand that animals have feelings and, more importantly, translate this into their daily lives and choices. Not only will our children benefit, but so, too, will future generations as we all negotiate the challenging and frustrating path through the Anthropocene.

I'm an optimist and really believe that with hard work, diligence, and courage we can right many of the wrongs that animals suffer at our hands. There are many wonderful people working in a myriad of ways, large and small, to make the lives of animals better. Some efforts are very public, and some are private, but together they help to realize a peaceable kingdom here on Earth, in which all beings are blanketed in a seamless tapestry of compassion and love. Surely, no one can argue that a world with more respect, compassion, and love would not be a better place in which to live and to raise all of our children. My message is a forward-looking one of hope. We must follow our dreams.

All I ask is that you reflect on how you can make the world a better place; chiefly, how you can contribute to making the lives of animals better. Do this when you're alone, away from others, so that you can feel free to look deeply and assess your current habits and actions absent peer or any other sort of pressure. It's always a sobering experience to try to view ourselves as we really are. In this case, ask yourself, how do your current actions affect other animals, and what can you do differently to care for animals better? Even when a situation is beyond my ability to change it, I make a point of apologizing to each and every individual animal who finds themselves being subjected to inhumane treatment. I believe that even just the expression of compassion

Afterword: Compassion and Justice for All

can make a positive difference in the life of someone who is suffering. Silence is the enemy of social change.

We owe it to all individual animals to make every attempt to come to a greater understanding and appreciation for who they are in their world and in ours. We must make kind and humane choices. There's nothing "radical" about treating other animals respectfully. It's a matter of decency and dignity because they are fully worthy of living the best lives possible. There's also nothing to fear and much to gain by being open to such deep and reciprocal interactions with other animals. Animals have in fact taught me a great deal: about responsibility, compassion, caring, forgiveness, and the value of deep friendship and love. Animals generously share their hearts with us, and I want to do the same. Animals respond to us because we are feeling and passionate beings, and we embrace them for the same reasons.

Emotions are the gifts of our ancestors. We have them and so do other animals. Emotions can reconnect us to the world at large and help us truly coexist with other animals. We're all on this journey together and every single individual matters. When we help other animals, we also help ourselves. By allowing ourselves to emotionally connect with all life, we can make life better for everyone — a win-win for all. We must never forget this.

Acknowledgments

First, I thank all the wonderful animal beings I've been lucky enough to know in a wide variety of situations. Their willingness to share their lives and their passions with me was instrumental in my deciding that I wanted to come to terms with their emotions and their worldviews. Numerous canine buddies have patiently listened to me talk about their feelings and made me a better dog. So, too, have the animals with whom I shared my home range when I lived in the mountains outside of Boulder.

Many of the people mentioned in the first edition of this book continue to influence how I see other animals, and I humbly thank all of the people who have sent me stories and continue to do so about animals in their lives. I couldn't have updated this book without their contributions. Jessica Pierce will clearly see some of her and our ideas sprinkled throughout this book, and I cannot put into words how much I owe to her. I also learned a lot from Colin Allen, Dale Jamieson, and Sarah Bexell. I also deeply thank Jane Goodall for updating her previous foreword and for many discussions about the emotional lives of animals, animal protection, and the importance of maintaining hope. Koen Margodt, with whom I co-chair the ethics committee of the Jane Goodall Institute, also has taught me some valuable lessons about animal ethics. Colorado's first gentleman Marlon H. Reis is always there to talk about "all things animals" and animal protection, and I so value his willingness to chat for hours on end about how to make our world a better place for all nonhuman and human animals. And to all my cycling friends who have to listen to me marvel about the animals we see on our rides, I thank you for tolerating my obsessions and for asking great questions about what's happening when we see nonhumans doing what comes naturally. I also thank the indefatigable Betty Moss for continually sending me the latest and greatest news about all things animals. Others, too, including Mary Lewis and Paul Paquet, don't hesitate to fill my inbox with articles I otherwise never would have seen — a huge thanks to them as well.

When I write about and observe animals, I always remember Boniface Zakaria, my guide in Tanzania, who spotted a tiny chameleon on a blade of grass while driving at fifteen miles an hour in Serengeti National Park — a little speck I couldn't see until I was standing six inches away. This made me fully realize how much keen observers of animal behavior can miss even when we're giving our full attention to what we're doing. Thank you, Boniface, for this humbling lesson.

At New World Library, Jason Gardner, with whom I've done five other books, continues to support my work, and I am forever grateful to him for his friendship and keen eyes. Monique Muhlenkamp continues to be an awesome publicist, and Ami Parkerson, Danielle Galat, Joel Prins, Kristen Cashman, Tracy Cunningham, Tona Pearce-Myers, Munro Magruder, and outside proofreader Tanya Fox are always there when I need them. Jeff Campbell once again did a wonderful job of copyediting the manuscript. Working with the entire crew at New World Library and Jeff over the course of many years has been a blessing.

Endnotes

This section contains information about the sources I used in writing this book. I've also included websites for summaries of many of the technical papers.

Foreword

For more information about the Jane Goodall Institute, please visit janegoodall.org.

p. xii   *It also looks like they can have nightmares*: Carolyn Wilke, "Is This Octopus Having a Nightmare?," *New York Times*, May 25, 2023, https://www.nytimes.com/2023/05/25/science/octopus-nightmare-dream.html.

Introduction: The Gift of Animal Emotions

p. 1   *"We like to see ourselves as special"*: Frans de Waal, "Your Dog Feels as Guilty as She Looks," *New York Times*, March 8, 2019, https://www.nytimes.com/2019/03/08/opinion/sunday/emotions-animals-humans.html.

p. 1   *in part through my Psychology Today blog*: Marc Bekoff, *Animal Emotions* (blog), *Psychology Today*, https://www.psychologytoday.com/us/blog/animal-emotions. The site has gotten more than 10 million views, and it's good reflection that numerous people really want to learn about the cognitive, emotional, and moral lives of animals.

p. 3   *I have always argued that stripping animals of emotions*: Marc Bekoff, "Stripping Animals of Emotions Is 'Anti-Scientific & Dumb,'" *Psychology Today*, March 9, 2019, https://www

p. 3   .psychologytoday.com/us/blog/animal-emotions/201903/stripping-animals-emotions-is-anti-scientific-dumb.

p. 3   *"For the longest time, science has depicted animals"*: Frans de Waal, "Your Dog Feels as Guilty as She Looks," *New York Times*, March 8, 2019, https://www.nytimes.com/2019/03/08/opinion/sunday/emotions-animals-humans.html.

p. 3   *Later, de Waal wrote, "I now believe"*: Marc Bekoff, "Animal Emotions: What Must We Do with What We Know?," *Psychology Today*, March 13, 2019, https://www.psychologytoday.com/us/blog/animal-emotions/201903/animal-emotions-what-must-we-do-what-we-know.

p. 3   *Their declaration proposed that the real question*: Among the few remaining skeptics is Oxford University professor Marian Dawkins, who has written widely on animal welfare. She claims we really are in a fog concerning human and nonhuman animal consciousness. For example, in *Why Animals Matter* (page 177), she writes, "It is much, much better *for animals* if we remain skeptical and agnostic [about consciousness].... Militantly agnostic if necessary, because this keeps alive the possibility that a large number of species have some sort of conscious experiences.... For all we know, many animals, not just the clever ones and not just the overtly emotional ones, also have conscious experiences." I'm frankly confused by this and not sure what to make of it. I call it "Dawkins' dangerous idea."

p. 4   *"Animals in Spain will no longer be considered"*: Xosé Hermida and Esther Sánchez, "Spain Approves New Law Recognizing Animals as 'Sentient Beings.'" *El País*, December 3, 2021, https://english.elpais.com/society/2021-12-03/spain-approves-new-law-recognizing-animals-as-sentient-beings.html.

p. 4   *In April 2022, Ecuador became the first country*: Tessa Koumoundouros, "The First Country in the World Has Given Legal Rights to Individual Wild Animals," *Science Alert*, April 2, 2022, https://www.sciencealert.com/the-first-country-in-the-world-has-given-legal-rights-to-individual-wild-animals.

p. 6   *Recent research has shown that elephants*: Franziska Hörner et al., "Long-Term Olfactory Memory in African Elephants," *Animals* 13, no. 4 (2023), https://www.mdpi.com/2076-2615/13/4/679.

p. 6   *they possess what psychologists call emotional intelligence*: Colleen Dougherty, "Emotional Intelligence: Animals May Have the Corner on This Skill," *Valencia County News-Bulletin*, June 16, 2022, https://news-bulletin.com/emotional-intelligence-animals-may-have-the-corner-on-this-skill.

p. 6   *in 2023, two rescue dogs, a male named Heart*: Alice Gibbs, "Loyal Dog Refuses to Leave Canine's Side after Being Hit by Car: 'Grieving.'" *Newsweek*, January 3, 2023, https://www.newsweek.com/loyal-dog-refuses-leave-canines-side-hit-car-1770848.

p. 6   *Such was the case with Aochan, a rat snake, who befriended*: For this and other "odd couple" stories, see Jennifer Holland's books *Unlikely Friendships* (New York: Workman Publishing, 2011) and *Unlikely Loves* (New York: Workman Publishing, 2013). The other stories here are from Lauren Lewis, "Bottlenose Dolphins & Harbour Porpoise Form an Unlikely Friendship While Swimming & Playing in the Waters off Cornwall, England," *World Animal News*, March 13, 2023, https://worldanimalnews.com/bottlenose-dolphins-harbour-porpoise-form-a-rare-bond-while-swimming-and-playing-in-the-waters-of-cornwall-england; and Margaret Renkl, "There Are Stories of Animals Being Quirky, and Murphy the Eagle Is One," *New York Times*, May 1, 2023, https://www.nytimes.com/2023/05/01/opinion/murphy-eagle-rock-baby.html.

p. 7   *Books like* Daisy to the Rescue *by Jeff Campbell*: Marc Bekoff, "Animal Heroes: *Daisy to* the Rescue Celebrates Compassion," *Psychology Today*, November 6, 2014, https://www.psychologytoday.com/au/blog/animal-emotions/201411/animal-heroes-daisy-to-the-rescue-celebrates-compassion; and Marc Bekoff, "'When Animals Rescue': Reflections on Kindness and Morality," *Psychology Today*, April 30, 2021, https://www.psychologytoday.com/us/blog/animal-emotions/202104/when-animals-rescue-reflections-kindness-and-morality.

p. 7   *As Frans de Waal notes, if we deny sentience*: Franz de Waal, *Mama's Last Hug* (New York: W. W. Norton, 2019), 269.

p. 8   *some scientists think AI might provide the communication breakthrough*: Sophie Bushwick, "How Scientists Are Using AI to Talk to Animals," *Scientific American*, February 7, 2023, https://www.scientificamerican.com/article/how-scientists-are-using-ai-to-talk-to-animals.

p. 9   *My hope is that the jury is still out on whether*: Marc Bekoff, "Is Human Intelligence a Gift or a Burden?," *Psychology Today*, August 9, 2022, https://www.psychologytoday.com/us/blog/animal-emotions/202208/is-human-intelligence-gift-or-burden.

p. 9   *In 2023, a Spanish business proposed creating an octopus*: Charles R. Davis, "A Spanish Firm Wants to Kill One Million Octopuses a Year Using 'Ice Slurry' Baths at First-Ever Factory Farm," *Business Insider*, March 16, 2023, https://www.businessinsider.com/first-ever-proposed-octopus-farm-sparks-concern-over-conditions-2023-3.

p. 9   *The same year, at a Spanish hunting fair, fifty-four outfitters*: Lauren Lewis, "Sickening Trophy Hunting Event in Spain Auctioned Off Hunts of Lions, Rhinos, and Polar Bears, Among Other Endangered Species," *World Animal News*, April 3, 2023, https://worldanimalnews.com/sickening-trophy-hunting-event-in-spain-promoted-the-killing-of-rhinos-polar-bears-elephants-among-other-endangered-species.

p. 10   *I'm pleased that more attention is being paid to using the word* dignity: For instance, see Melanie Challenger, ed., *Animal Dignity* (London: Bloomsbury Academic, 2023); and Lauren Krauze, "Restoring Dignity to Our Animal Kin," *Tricycle*, February 28, 2023, https://tricycle.org/article/amanda-stronza/?fbclid=IwAR2_trYMhe-fqpGVkeg7zO0A7Ih6dHaTTEPaf-J-BguLe24nJM1NgODsuGk.

p. 10   *"the world is still a better place because those rights exist"*: Graham Lawton, "The Push to Grant Legal Rights to Nature Is Gaining Momentum," *New Scientist*, February 8, 2023, https://www.newscientist.com/article/mg25734250-500-the-push-to-grant-legal-rights-to-nature-is-gaining-momentum.

p. 10   *Researchers have learned that by protecting certain groups*: Christa Lesté-Lasserre, "Restoring Just Nine Groups of Animals Could Help Combat Global Warming," *New Scientist*, March 27, 2023, https://www.newscientist.com/article/2366378-restoring-just-nine-groups-of-animals-could-help-combat-global-warming.

p. 11   *Personally, I have always felt that dogs are the best species*: Marc Bekoff, "'Everyone Wants a Lost Dog Found,' Bridging the Empathy Gap," *Psychology Today*, May 9, 2018, https://www.psychologytoday.com/us/blog/animal-emotions/201805/everyone-wants-lost-dog-found-bridging-the-empathy-gap. Research shows that children who spend a lot of time with dogs are more likely to say they would save a dog over a person; see Alice Klein, "Young Children Value the Lives of Animals More Than Adults Do," *New Scientist*, May 14, 2023, https://www.newscientist.com/article/2372041-young-children-value-the-lives-of-animals-more-than-adults-do.

Chapter 1: The Indisputable Case for Animal Emotions and Why They Matter

p. 13   *"Anthropocentrism is at the root of all abuse"*: Marc Bekoff and Koen Margodt, eds., *Jane Goodall at 90: Celebrating an Astonishing Lifetime of Science, Advocacy, Humanitarianism, Hope, and Peace* (Berlin, MD: Salt Water Media, 2024).

p. 14   *Simply put, science now confirms that what we see outside*: Patricia McConnell also makes this point in her book *For the Love of a Dog: Understanding Emotion in You and Your Friend* (New York: Ballantine Books, 2006).

p. 15   *"Crows and ravens routinely gather around the dead"*: John Marzluff and Tony Angell, *Gifts of the Crow* (New York: Atria, 2013), 138.

p. 15   *"I heard a sound at the window of the cottage"*: Hasan Jasim, "Yesterday I Witnessed Something That Took My Breath Away," *Hasan Jasim* (blog), accessed September 26, 2023, https://hasanjasim.online/yesterday-i-witnessed-something-that-took-my-breath-away.

p. 16   *My encounters in Africa were deeply spiritual*: In 2023, I was thrilled to learn that in the southern state of Kerala, an Indian temple was going cruelty-free and began using a life-sized robotic elephant in its ceremonies because of welfare concerns for live elephants; see Sana Noor Haq, "Indian Temple Goes 'Cruelty-Free' with Life-Sized Robotic Elephant," *CNN Travel*, February 28, 2023, https://www.cnn.com/travel/article/india-temple-elephant-robot-intl-scli/index.html.

p. 16   *In October 2006, in a small village in eastern India*: Reuters, "Indians Flee as Elephants Search for Dead Friend," *NBC News*, October 10, 2006, https://www.nbcnews.com/id/wbna15204018.

p. 17   *"Emotions originate as sensations in the body"*: Rachel Allyn, "The Important Difference between Emotions and Feelings," *Psychology Today*, February 23, 2022, https://www.psychologytoday.com/us/blog/the-pleasure-is-all-yours/202202/the-important-difference-between-emotions-and-feelings.

p. 17   *Charles Darwin, the first scientist to study animal emotions*: The complete works of Darwin, along with details about Darwin's work on emotions, can be found at http://darwin-online.org.uk.

p. 18   *"While I feel confident that elephants feel"*: Joyce Poole, "An Exploration of a Commonality between Ourselves and Elephants," *Etica & Animali* (September 1998): 85–110.

p. 20   *Most people with dogs are familiar with zoomies*: For more about zoomies and play, see Marc Bekoff, "Dogs Do Zoomies Because It's Fun," *Psychology Today*, February 24, 2023, https://www.psychologytoday.com/us/blog/animal-emotions/202302/dogs-do-zoomies-because-its-fun; and Marc Bekoff, "Do Animals Play for the Hell of It? Watch This Fox," *Psychology Today*, December 5, 2014, https://www.psychologytoday.com/us/blog/animal-emotions/201412/do-animals-play-the-hell-it-watch-fox.

p. 21   *"The owner says without special intonation"*: Konrad Lorenz, *Man Meets Dog* (New York: Routledge, 1954/2002).

p. 21   *as Alexandra Horowitz and I have argued*: Alexandra C. Horowitz and Marc Bekoff, "Naturalizing Anthropomorphism: Behavioral Prompts to Our Humanizing of Animals," *Anthrozoös* (2007).

p. 21   *One example of how scientists accept the universality*: J. P. Webster et al., "Parasites as Causative Agents of Human Affective Disorders? The Impact of Anti-psychotic, Mood-stabilizer and Anti-parasite Medication on *Toxoplasma gondii*'s Ability to Alter Host Behaviour," *Proceedings of the Royal Society of London, Series B: Biological Sciences* 273 (2006):

1023–30; and "Antipsychotic Drug Lessens Sick Rats' Suicidal Tendencies," *New Scientist*, January 25, 2006, http://www.newscientist.com/channel/health/mg18925365.000.html.

p. 22   *I read a similar story about two grizzly bear cubs*: Doug O'Harra, "Russian River Orphans Stick Together," *Anchorage Daily News*, September 23, 2005.

p. 22   *There's also a story of a troop of about a hundred rhesus monkeys*: Marc Bekoff, "The Depths of Feeling," *BBC Wildlife*, July 2002.

p. 22   *In one classic study, a hungry rhesus monkey*: S. Wechlin, J. H. Masserman, and W. Terris Jr., "Shock to a Conspecific as an Aversive Stimulus," *Psychonomic Science* 1 (1964): 17–18.

p. 22   *there is a more recent scientific study on empathy in mice*: D. J. Langford et al., "Social Modulation of Pain as Evidence for Empathy in Mice," *Science* 312 (2006): 1967–70; and Ishani Ganguli, "Mice Show Evidence of Empathy," *The Scientist*, June 30, 2006.

p. 22   *Two other studies showed that rats would free trapped rats*: Marc Bekoff, "Empathic Rats Save Drowning Pals Rather Than Eat Chocolate," *Psychology Today*, May 13, 2015, https://www.psychologytoday.com/us/blog/animal-emotions/201505/empathic-rats-save-drowning-pals-rather-eat-chocolate.

p. 23   *A wonderful example of cross-species empathy and compassion*: Amelia Neath, "Crocodiles Push Stray Dog to Safety After It Falls into Danger-Filled River," *Independent*, September 29, 2023, https://www.independent.co.uk/news/science/stray-dog-safety-crocodiles-india-b2420758.html.

p. 23   *"The reluctance of contemporary philosophers"*: Dale Jamieson, "Science, Knowledge, and Animal Minds," *Proceedings of the Aristotelian Society* 98 (1998): 79–102.

p. 24   *A few animals, such as chimpanzees, dolphins, elephants*: See Andrew Bridges, "Mirror Test Implies Elephants Self-Aware," *Associated Press*, October 31, 2006; Joshua Plotnik, Frans de Waal, and Diana Reiss, "Self-Recognition in an Asian Elephant," *Proceedings of the National Academy of Sciences* 45 (November 7, 2006); and Betsy Mason, "Fish Can Recognize Themselves in Photos, Further Evidence They May Be Self-Aware," *Science News*, February 6, 2023, https://www.sciencenews.org/article/fish-recognize-photo-self-aware.

p. 24   *Australian biologist Gisela Kaplan learned that birds*: Marc Bekoff, "Brainy Birds Are More Socially Playful Than the Less Endowed," *Psychology Today*, December 1, 2020, https://www.psychologytoday.com/us/blog/animal-emotions/202012/brainy-birds-are-more-socially-playful-the-less-endowed.

p. 24   *"The human animal is the only animal that thinks"*: Daniel Gilbert, *Stumbling on Happiness* (New York: Alfred A. Knopf, 2006), 4.

p. 25   *Another example is Gerald Hüther's claim*: Gerald Hüther, *The Compassionate Brain: How Empathy Creates Intelligence* (Boston: Trumpeter, 2006), 114.

p. 26   *Bees have small but stunning brains and are surely sentient*: For more about bees, see Annette McGivney, "'Bees Are Sentient': Inside the Stunning Brains of Nature's Hardest Workers," *The Guardian*, April 2, 2023, https://www.theguardian.com/environment/2023/apr/02/bees-intelligence-minds-pollination; Jason Castro, "Bees Appear to Experience Moods," *Scientific American*, January 1, 2012, https://www.scientificamerican.com/article/the-secret-inner-life-of-bees; and Virginia Morell, "Bees May Feel Pain," *Science*, July 26, 2022, https://www.science.org/content/article/bees-may-feel-pain.

p. 26   *In a detailed 2022 review called "Can Insects Feel Pain?"*: Matilda Gibbons et al., "Can Insects Feel Pain? A Review of the Neural and Behavioural Evidence," *Advances in Insect Physiology* 63 (2022): 155–229, https://www.sciencedirect.com/science/article/abs/pii/S0065280622000170.

p. 26   *Biologist Meghan Barrett and her colleagues are deeply concerned*: Meghan Barrett, Bob

Fischer, and Stephen Buchmann, "Informing Policy and Practice on Insect Pollinator Declines: Tensions between Conservation and Animal Welfare," *Frontiers in Ecology and Evolution* 10 (January 6, 2023), https://www.frontiersin.org/articles/10.3389fevo.2022.1071251/full.

p. 26   *This essay laid out eight criteria to assess sentience*: Jonathan Birch et al., "Review of the Evidence of Sentience in Cephalopod Molluscs and Decapod Crustaceans," LSE Consulting, LSE Enterprise Ltd., London School of Economics and Political Science, November 2021, https://www.wellbeingintlstudiesrepository.org/cgi/viewcontent.cgi?article=1001&context=af_gen. The authors write: "To move from the individual criteria to an overall judgement, we use an approximate grading scheme. In our scheme, high or very high confidence that an animal satisfies 7 or more of the criteria amounts to very strong evidence of sentience. High or very high confidence that an animal satisfies 5 or more criteria amounts to strong evidence of sentience, and high or very high confidence that an animal satisfies 3 or more criteria amounts to substantial evidence of sentience."

p. 27   *Fishes are also cunning, deceitful, and display cultural traditions*: For more on the cognitive and emotional lives of fishes, see Jonathan Balcombe's *What a Fish Knows: The Inner Lives of Our Underwater Cousins* (New York: Scientific American / Farrar, Straus and Giroux, 2016); Marc Bekoff, "Why Fishes Matter: Their Rich Cognitive and Emotional Lives," *Psychology Today*, August 31, 2020, https://www.psychologytoday.com/us/blog/animal-emotions/202008/why-fishes-matter-their-rich-cognitive-and-emotional-lives; and Marc Bekoff, "A Tribute to Dr. Victoria Braithwaite and Sentient Fishes," *Psychology Today*, November 4, 2019, https://www.psychologytoday.com/us/blog/animal-emotions/201911/tribute-dr-victoria-braithwaite-and-sentient-fishes.

p. 27   *"I have argued that there is as much evidence that fish"*: Victoria Braithwaite, *Do Fish Feel Pain?* (New York: Oxford University Press, 2010), 153.

p. 28   *Some suggest that plants do feel pain and may even "cry"*: Emma Marris, "Stressed Plants 'Cry' — and Some Animals Can Probably Hear Them," *Nature*, March 30, 2023, https://www.nature.com/articles/d41586-023-00890-9.

p. 28   *Likewise, just as ample data show that many people find*: Marc Bekoff, "Why Trees and People Form Mutually Beneficial Relationships," *Psychology Today*, February 2, 2023, https://www.psychologytoday.com/us/blog/animal-emotions/202301/why-trees-and-people-form-mutually-beneficial-relationships.

p. 28   *many scientists studying plants have moved on from asking* if *plants*: For more on plants, see some of my *Psychology Today* articles: "What's It Like to Be a Plant? An Interview with 'Planta Sapiens' Author," September 23, 2022, https://www.psychologytoday.com/us/blog/animal-emotions/202209/what-s-it-be-plant-interview-planta-sapiens-author; "The Inner Lives of Plants: Cognition, Sentience, and Ethics," September 10, 2022, https://www.psychologytoday.com/us/blog/animal-emotions/202209/the-inner-lives-plants-cognition-sentience-and-ethics; "Are Plants Intelligent?," December 16, 2021, https://www.psychologytoday.com/us/blog/animal-emotions/202112/are-plants-intelligent; and "Smarty Plants: Research Shows They Think, Feel, and Learn," December 9, 2014, https://www.psychologytoday.com/us/blog/animal-emotions/201412/smarty-plants-research-shows-they-think-feel-and-learn.

p. 28   *Shared emotions and their glue-like adhesive power*: There is even a field of study, called anthrozoology, devoted to human and nonhuman animal interactions. For more information,

Endnotes

check out the website for the International Society for Anthrozoology (ISAZ) at https://isaz.net.

p. 28 *In 2022, around 70 percent of US households had a least one*: "Pet Ownership Statistics in 2022 — 70 Fun Fur Facts," Lemonade.com, accessed September 26, 2023, https://www.lemonade.com/pet/explained/pet-ownership-statistics/#US.

p. 29 *One study showed that a visit from a friendly pup*: Ed Sussman, "AHA: Cardio, the Canine Heart Dog, Is a Friend Indeed," *MedPage Today*, November 17, 2005.

p. 29 *That said, animals are not necessarily a global panacea*: The details in this paragraph are from Marc Bekoff, "Do Pets Really Unconditionally Love and Unwind Us?," *Psychology Today*, June 15, 2020, https://www.psychologytoday.com/us/blog/animal-emotions/202006/do-pets-really-unconditionally-love-and-unwind-us; Hal Herzog, "Can Pets Relieve Loneliness in the Age of Coronavirus?," *Psychology Today*, April 13, 2020, https://www.psychologytoday.com/us/blog/animals-and-us/202004/can-pets-relieve-loneliness-in-the-age-coronavirus; and Hal Herzog, "The Sad Truth About Pet Ownership and Depression," *Psychology Today*, December 3, 2019, https://www.psychologytoday.com/us/blog/animals-and-us/201912/the-sad-truth-about-pet-ownership-and-depression.

p. 30 *For example, three lions in Ethiopia rescued a twelve-year-old girl*: Anthony Mitchell, "Three Lions Save Girl, 12, from Kidnap Gang," *The Scotsman*, June 22, 2005.

p. 30 *in New Zealand, a pod of dolphins circled protectively*: Reuters, "Dolphins Protect Swimmers from Sharks," *MSNBC*, November 23, 2004.

p. 30 *Some of the relationships that animals form are so improbable*: The lioness story is from Saba Douglas-Hamilton, "Heart of a Lioness," January 2002, https://sabadouglashamilton.com/heart-of-a-lioness; the story of Aochan is from Associated Press, "Hamster, Snake Best Friends at Tokyo Zoo," *MSNBC*, January 24, 2006.

p. 31 *After they were rejected by their parents on a UK farm, Larry the lamb*: Tom Burgess and Will Maule, "Sheep and Duck Become Unlikely Best Friends after They're Rejected by Their Parents," *Mirror*, January 1, 2023, https://www.mirror.co.uk/news/uk-news/sheep-duck-become-unlikely-best-28850294.

p. 31 *Jane Goodall's Roots & Shoots program*: For more information on Roots & Shoots, visit https://rootsandshoots.org.

p. 31 *It's estimated that around 40 percent of children*: Bill Strickland, "6 Surprising Benefits of Pets for Kids," *Parents*, September 12, 2023, https://www.parents.com/parenting/pets/kids/the-benefits-of-pets.

p. 32 *Research and studies show that living with dogs or cats*: "Pets and Children," American Academy of Child & Adolescent Psychiatry, January 2019, https://www.aacap.org/AACAP/Families_and_Youth/Facts_for_Families/FFF-Guide/Pets-And-Children-075.aspx; and Kristen Jacobson and Laura Chang, "Associations between Pet Ownership and Attitudes toward Pets with Youth Socioemotional Outcomes," *Frontiers in Psychology* 9 (2018), https://www.ncbi.nlm.nih.gov/pmc/articles/PMC6275470.

p. 32 *Pets can help victims of psychological abuse by teaching*: Lauren Book, "Therapy Animals Help Abuse Victims with Confidence, Trust, and Recovery," *Lauren's Kids* (blog), January 17, 2015; and Stephanie Johnston, "The Use of Animal-Assisted Interventions to Treat Child Victims of Sexual Abuse," master's thesis, Dominican University of California, May 2021, https://scholar.dominican.edu/counseling-psychology-masters-theses/3.

p. 32 *Perhaps the most heartbreaking story about a young girl*: María Luisa Paúl, "10 Years after Sandy Hook, Mom Builds Daughter's Dream Animal Sanctuary," *Washington Post*,

December 14, 2022, https://www.washingtonpost.com/nation/2022/12/14/hubbard-animal-sanctuary-sandy-hook.

p. 32 *In my classes at the Boulder County Jail*: For more information, see the following: Boulder Art Behind Bars, https://www.boulderjailart.com; Caitlin Rockett, "Roots and Shoots," *Boulder Weekly*, April 13, 2017, https://boulderweekly.com/news/roots-and-shoots; Marc Bekoff, "Inmates, Animals, and Art: Creative Expressions of Hope," *Psychology Today*, December 17, 2016, https://www.psychologytoday.com/intl/blog/animal-emotions/201612/inmates-animals-and-art-creative-expressions-hope; and Marc Bekoff, "Art Behind Bars: Animals, Compassion, Freedom, and Hope," *Psychology Today*, June 24, 2019, https://www.psychologytoday.com/us/blog/animal-emotions/201906/art-behind-bars-animals-compassion-freedom-and-hope.

p. 33 *"This is my spirit animal — the moon bear"*: Kyle Warner (guest essay), "The Healing Power of Animals: Moon Bear Has a Place," *Psychology Today*, May 19, 2019, https://www.psychologytoday.com/us/blog/animal-emotions/201905/the-healing-power-animals-moon-bear-has-place.

p. 34 *"More than any other species, we are the beneficiaries"*: R. J. Dolan, "Emotion, Cognition, and Behavior," *Science* 298 (2002): 1191–94.

p. 34 *There is no evidence that there are degrees of sentience*: Marc Bekoff, "'Are Chimps and Dogs More Sentient than Rats and Hamsters?,'" *Psychology Today*, September 1, 2022, https://www.psychologytoday.com/us/blog/animal-emotions/202209/are-chimps-and-dogs-more-sentient-rats-and-hamsters.

p. 36 *Compassionate conservation came of age around 2010*: Marc Bekoff and Dale Jamieson, "Ethics and the Study of Carnivores: Doing Science While Respecting Animals," in *Carnivore Behavior, Ecology, and Evolution*, ed. J. Gittleman (Ithaca, NY: Cornell University Press, 1996), 16–45.

p. 37 *As I mention, revising the US Federal Animal Welfare Act*: Marc Bekoff, "The Animal Welfare Act Claims Rats and Mice Are Not Animals," *Psychology Today*, September 25, 2016, https://www.psychologytoday.com/us/blog/animal-emotions/201609/the-animal-welfare-act-claims-rats-and-mice-are-not-animals.

p. 37 *with the help of the growing number of scientists*: For more information on these three scientists, see Speaking of Research, "The Double Life of Dr. Lawrence A. Hansen," accessed September 27, 2023, https://speakingofresearch.com/2013/08/16/the-double-life-of-dr-lawrence-a-hansen; John P. Gluck, *Voracious Science and Vulnerable Animals: A Primate Scientist's Ethical Journey* (Chicago: University of Chicago Press, 2016); Garet Lahvis, "Freefall into Darkness," *Aeon*, June 2, 2022, https://aeon.co/essays/what-do-caged-animals-really-tell-us-about-our-mental-lives.

p. 37 *Every man and living creature has the sacred right*: https://www.goodreads.com/quotes/39162-every-man-and-every-living-creature-has-a-sacred-right.

p. 37 *Pablo was a captive and mistreated chimpanzee*: The source for Pablo's story is J. D'Agnese, "An Embarrassment of Chimpanzees," *Discover* (May 2002): 42–49.

p. 37 *Jasper was a moon bear who was formerly kept in a crush cage*: Jill Robinson told the author Jasper's story via personal email, which is the source of the quotes.

p. 38 *The organization Animals Asia is leading the way*: Animals Asia, "Bear Bile Farming," accessed September 27, 2023, https://www.animalsasia.org/us/our-work/end-bear-bile-farming; and Animals Asia, "Animals Asia Rescues Seven Bears from Bile Farming Hotspot in Vietnam," July 7, 2022, https://www.animalsasia.org/us/media/news/news-archive/animals-asia-rescues-seven-bears-from-bile-farming-hotspot-in-vietnam.html.

Endnotes

Chapter 2: Cognitive Ethology

p. 41   *"What we believe* Equus asinus *most prefers"*: Michael Tobias and Jane Morrison, *Donkey: The Mystique of* Equus asinus (Tulsa, OK: Council Oak Books, 2006).

p. 42   *Interest in the emotional and mental capacities of animals*: Dale Jamieson and Marc Bekoff, "On Aims and Methods of Cognitive Ethology," *PSA: Proceedings of the Biennial Meeting of the Philosophy of Science Association* 1992 (1992): 110–24.

p. 43   *Research has shown that rats often take a moment*: Nicholas Wade, "Rats in a Maze Take a Moment to Remember, but in Reverse," *New York Times*, February 14, 2006, http://www.nytimes.com/2006/02/14/science/14rats.html.

p. 43   *birds form what Ashley Ward calls "clusterflocks"*: Ashley Ward, *The Social Lives of Animals* (New York: Basic Books, 2022).

p. 43   *Dairy cows live highly constrained, abused lives*: Marina A. G. von Keyserlingk et al., "Dairy Cows Value Access to Pasture as Highly as Fresh Feed," *Scientific Reports* 7 (March 23, 2017), https://doi.org/10.1038/srep44953.

p. 44   *"I formerly possessed a large dog"*: Charles Darwin, *The Expression of the Emotions in Man and Animals*, third edition (New York: Oxford University Press, 1872/1998), 58.

p. 45   *"There is no fundamental difference between man and the higher animals"*: Charles Darwin, *The Descent of Man and Selection in Relation to Sex* (New York: Random House, 1871/1936), 66.

p. 46   *Neuroimaging shows that when dogs are feeling jealous*: Peter Cook et al., "Jealousy in Dogs? Evidence from Brain Imaging," *Animal Sentience* 22, no. 1 (2018), https://www.wellbeingintlstudiesrepository.org/cgi/viewcontent.cgi?article=1319&context=animsent.

p. 46   *And the so-called "love hormone" oxytocin produced*: Markus MacGill, "What Is the Link between Love and Oxytocin?," *Medical News Today*, September 4, 2017, https://www.medicalnewstoday.com/articles/275795.

p. 46   *In 1973, Tinbergen, Konrad Lorenz, and Karl von Frisch won the Nobel Prize*: For more information about these ethologists (and others), see the following books: Richard W. Burkhardt Jr., *Patterns of Behavior: Konrad Lorenz, Niko Tinbergen, and the Founding of Ethology* (Chicago: University of Chicago Press, 2005); Hans Kruuk, *Niko's Nature: The Life of Niko Tinbergen and His Science of Animal Behavior* (New York: Oxford University Press, 2004); and I. Eibl-Eibesfeldt, *Ethology* (New York: Holt, Rinehart, & Winston, 1975).

p. 47   *in the 1990s, psychologist Gordon Burghardt added a fifth area*: Gordon M. Burghardt, "Amending Tinbergen: A Fifth Aim for Ethology," in *Anthropomorphism, Anecdotes, and Animals*, ed. R. W. Mitchell, N. S. Thompson, and H. L. Miles (Albany: State University of New York Press, 1997), 254–76, https://psycnet.apa.org/record/1996-98870-019.

p. 48   *"A greylag goose that has lost its partner shows"*: Marc Bekoff, "Grief in Animals: It's Arrogant to Think We're the Only Animals Who Mourn," *Psychology Today*, October 29, 2009, https://www.psychologytoday.com/us/blog/animal-emotions/200910/grief-in-animals-its-arrogant-think-were-the-only-animals-who-mourn.

p. 48   *in December 2022, the UK news shared the story of a goose*: James Liddell, "Gretel the Goose Dies of Heartbreak After 'Soulmate' Hansel Gets Trapped in Frozen Lake," *Mirror*, December 20, 2022, https://www.mirror.co.uk/news/uk-news/gretel-goose-dies-heartbreak-after-28777918.

p. 49   *In the 1990s, scientist Michel Cabanac discovered that reptiles*: Michel Cabanac, "Emotion and Phylogeny," *Journal of Consciousness Studies* 6 (1999): 176–90.

p. 49   *In fact, subsequent research clearly shows that reptiles*: For more on reptiles, see Marc Bekoff, "The Emotional Lives of Reptiles: Stress and Welfare," *Psychology Today*, March 26, 2013, https://www.psychologytoday.com/us/blog/animal-emotions/201303/the-emotional-lives

-reptiles-stress-and-welfare; Marc Bekoff, "Sentient Reptiles Experience Mammalian Emotion," *Psychology Today*, November 1, 2019, https://www.psychologytoday.com/us/blog/animal-emotions/201911/sentient-reptiles-experience-mammalian-emotions; Helen Lambert et al., "Given the Cold Shoulder: A Review of the Scientific Literature for Evidence of Reptile Sentience," *Animals* 9, no. 10 (2019): 821, https://www.mdpi.com/2076-2615/9/10/821/htm; and Frances Mao, "Australia: Scientists Find Clitorises on Female Snakes," *BBC News*, December 13, 2022, https://www.bbc.com/news/world-australia-63967778.

p. 49   *In an interview, one of the coauthors, reptile expert Gordon Burghardt*: Marc Bekoff, "The Secret Social and Emotional Lives of Reptiles," *Psychology Today*, June 21, 2021, https://www.psychologytoday.com/us/blog/animal-emotions/202106/the-secret-social-and-emotional-lives-reptiles.

p. 49   *In May 2023, Panama granted legal rights to sea turtles*: Associated Press, "In Panama, Legal Rights Given to Sea Turtles, Boosting the 'Rights of Nature' Movement," *US News*, May 24, 2023, https://www.usnews.com/news/us/articles/2023-05-24/in-panama-legal-rights-given-to-sea-turtles-boosting-the-rights-of-nature-movement.

p. 49   *Another important recent book about reptiles*: Marc Bekoff, "Concerns Over the Well-Being of Sentient, Emotional Reptiles," *Psychology Today*, January 31, 2023, https://www.psychologytoday.com/us/blog/animal-emotions/202301/concerns-over-the-well-being-of-sentient-emotional-reptiles.

p. 52   *subsequent controlled experiments on lab dogs have confirmed*: Julie Hecht, "Do Dogs Know Themselves?," *Scientific American*, August 23, 2017, https://blogs.scientificamerican.com/dog-spies/do-dogs-know-themselves.

Chapter 3: Beastly Passions

p. 55   *"Much of chimpanzees' nonverbal communication is similar"*: Jane Goodall and Ray Greek, "The Sad Lot of Lab Chimps," *Boston Globe*, February 17, 2006.

p. 56   *Research by Sam Gosling and his colleagues has shown*: For more on Gosling's research, consult the following: Sam Gosling, "From Mice to Men: What Can We Learn about Personality from Animal Research?," *Psychological Bulletin* 127 (2001): 45–86; Sam Gosling and O. P. John, "Personality Dimensions in Non-human Animals: A Cross-species Review," *Current Directions in Psychological Science* 8 (1999): 69–75; Sam Gosling, P. J. Rentfrow, and W. B. Swann Jr., "A Very Brief Measure of the Big Five Personality Domains," *Journal of Research in Personality* 37 (2003): 504–28; and Sam Gosling and S. Vazire, "Are We Barking Up the Right Tree? Evaluating a Comparative Approach to Personality," *Journal of Research in Personality* 36 (2002): 607–14.

p. 56   *As it turns out, studies of the wolves in Yellowstone*: Connor J. Meyer et al., "Parasitic Infection Increases Risk-Taking in a Social, Intermediate Host Carnivore," *Communications Biology* 5, no. 1180 (2022), https://www.nature.com/articles/s42003-022-04122-0; Jake Buehler, "A Parasite Makes Wolves More Likely to Become Pack Leaders," *Science News*, December 6, 2022, https://www.sciencenews.org/article/wolves-parasite-pack-leader-toxoplasma-gondii; and Lina Zeldovich, "What Makes Us Bold," *Nautilus*, March 3, 2023, https://nautil.us/what-makes-us-bold-264300.

p. 56   *"I think people are just starting to really appreciate"*: Marc Bekoff, "Wolves with a Parasite Become More Daring, Study Shows," *Psychology Today*, January 15, 2023, https://www.psychologytoday.com/us/blog/animal-emotions/202301/wolves-with-a-parasite-become-more-daring-study-shows.

Endnotes

p. 57 *Research has shown that toxoplasmosis affects humans*: Toni Rodriguez, "Common Parasite Linked to Personality Changes," *Scientific American*, September 1, 2012, https://www.scientificamerican.com/article/common-parasite-linked-to-personality-changes.

p. 57 *Among wild elephants, older females run the show*: Lesley Evans Ogden, "What Elephants Can Teach Us about the Importance of Female Leadership," *Washington Post*, January 27, 2014, https://www.washingtonpost.com/national/health-science/what-elephants-can-teach-us-about-the-importance-of-female-leadership/2014/01/27/32db3f5e-7eeb-11e3-95c6-0a7aa80874bc_story.html; Roohi Narula, "A Trunk of Knowledge: What Female Elephants Can Teach Us about Forming the Ultimate Girl Gang!," Wildlife SOS, September 6, 2021, https://wildlifesos.org/elephant/a-trunk-of-knowledge-what-female-elephants-can-teach-us-about-forming-the-ultimate-girl-gang; and Marc Bekoff, "Age Before Beauty: Older Elephant Matriarchs Know What's Best," *Psychology Today*, April 11, 2011, https://www.psychologytoday.com/au/blog/animal-emotions/201104/age-beauty-older-elephant-matriarchs-know-whats-best.

p. 57 *Females are also very important in influencing the behavior*: Marc Bekoff, "The Power and Legacy of Yellowstone's Alpha Female Wolf 06," *Psychology Today*, November 8, 2022, https://www.psychologytoday.com/us/blog/animal-emotions/202211/the-power-and-legacy-yellowstones-alpha-female-wolf-06; see also Rick McIntyre, *The Alpha Female Wolf: The Fierce Legacy of Yellowstone's 06* (Vancouver, Canada: Greystone Books, 2022).

p. 57 *In her book* Bitch: On the Female of the Species: Marc Bekoff, "'Bitch' Repositions Female Animals to Where They Belong," *Psychology Today*, June 14, 2022, https://www.psychologytoday.com/us/blog/animal-emotions/202206/bitch-repositions-female-animals-where-they-belong.

p. 58 *Researchers have studied this and come to the same conclusion*: Extensive research has been conducted by Françoise Wemelsfelder and her colleagues. See the following: Françoise Wemelsfelder et al., "The Spontaneous Qualitative Assessment of Behavioural Expressions in Pigs: First Explorations of a Novel Methodology for Integrative Animal Welfare Measurement," *Applied Animal Behaviour Science* 67 (2000): 193–215; Françoise Wemelsfelder and A. B. Lawrence, "Qualitative Assessment of Animal Behaviour as an On-Farm Welfare-Monitoring Tool," *Acta Agriculturae Scandinavica* 30 (2001): 21–25 (supplement); Françoise Wemelsfelder and M. Farish, "Qualitative Categories for the Interpretation of Sheep Welfare: A Review," *Animal Welfare* 13 (2004): 261–68; Françoise Wemelsfelder et al., "Assessing the 'Whole Animal': A Free-Choice-Profiling Approach," *Animal Behaviour* 62 (2001): 209–20; and R. E. Anderson, J. Ryon, and J. C. Fentress, "Human Perception of Friendly and Agonistic Wolf Interactions," *Aggressive Behavior* 17 (1991): 58.

p. 58 *"He is a hot-blooded, 30-year-old male"*: "The Word: Musth," *New Scientist*, February 8, 2006, http://www.newscientist.com/channel/sex/mg18925381.900.html.

p. 59 *For instance, it's been determined that Leonardo da Vinci's* Mona Lisa: "Mona Lisa 'Happy,' Computer Finds," *BBC News*, December 15, 2005, http://news.bbc.co.uk/1/hi/entertainment/4530650.stm; and Associated Press, "Was Mona Lisa Pregnant When She Posed?," *MSNBC*, September 27, 2006.

p. 60 *"The elephants came at all times of the year"*: M. J. Owens and Delia Owens, *Secrets of the Savanna* (Boston: Houghton Mifflin, 2006), 2.

p. 60 *"But most disturbing of all, in Blue's large brown eyes"*: Alice Walker, "Am I Blue?" in *Living by the Word* (New York: Harcourt Brace Jovanovich, 1988), 8.

p. 60 *"The last time I looked into Five's eyes"*: Doug Smith, "Meet Five, Nine, and Fourteen, Yellowstone's Heroine Wolves," *Wildlife Conservation* (February 2005), 33.

p. 61 *"It was those eyes more than anything"*: Charles Siebert, "The Animal Self," *New York Times Magazine*, January 22, 2006, http://www.nytimes.com/2006/01/22/magazine/22animal.html.

p. 61 *For instance, eyes played a central role in the well-known story*: The story of Rick Swope and JoJo can be found in Jane Goodall, *Reason for Hope: A Spiritual Journey* (New York: Warner Books, 2000). See also Jane Goodall, "Essays on Science and Society," *Science* 5397 (December 18, 1998): 2184–85, http://www.sciencemag.org/cgi/content/full/282/5397/2184.

p. 61 *In 2005, three men near my hometown of Boulder*: This story appeared in the *Boulder Camera*, February 1, 2005.

p. 61 *Similarly, a study looking at fear in humans*: P. Vuilleumier, "Staring Fear in the Face," *Nature* 433 (2005): 22–23.

p. 61 *In December 2005, a fifty-foot, fifty-ton female humpback whale*: Peter Fimrite, "Daring Rescue of Whale off Farallones," *San Francisco Chronicle*, December 14, 2005, https://www.sfgate.com/bayarea/article/Daring-rescue-of-whale-off-Farallones-Humpback-2557146.php.

p. 62 *In a follow-up story in* Reader's Digest, *Frances Gulland*: Anita Bartholomew, "Whale of a Rescue," *Reader's Digest*, May 2006.

p. 62 *While there have been others, one such occasion was in April 2023*: Stephen Messenger, "Whale Approaches Boat Captain to Ask Him to Do Her a Helpful Favor," *The Dodo*, April 7, 2023, https://www.thedodo.com/amphtml/daily-dodo/whale-approaches-boat-captain-to-ask-him-to-do-her-a-helpful-favor.

p. 63 *"the pursuit of enjoyment is a primary emotion in our lives"*: Daniel Goleman, *Destructive Emotions: A Scientific Dialogue with the Dalai Lama* (New York: Bantam, 2004), 200.

p. 63 *One researcher told of watching a female chimpanzee*: Douglas Starr, "Animal Passions," *Psychology Today*, March/April 2006, 98.

p. 63 *According to Rosamund Young in her book* The Secret Life: Rosamund Young, *The Secret Life of Cows* (Preston, UK: Farming Books and Videos Ltd., 2005).

p. 64 *and perhaps so, too, do bumblebees*: Hiruni Samadi Galpayage Dona et al., "Do Bumble Bees Play?," *Animal Behaviour* 194 (2022): 239–53; Marc Bekoff, "Bumble Bees Play with Balls and May Even Enjoy It," *Psychology Today*, November 7, 2022, https://www.psychologytoday.com/us/blog/animal-emotions/202211/bumble-bees-play-balls-and-may-even-enjoy-it.

p. 64 *Dr. Karen Davis, president of United Poultry Concerns*: Karen Davis, "The Social Life of Chickens," United Poultry Concerns, accessed September 27, 2023, https://www.upc-online.org/thinking/social_life_of_chickens.html. Another wonderful love letter to chickens is Tove Danovich, *Under the Henfluence* (Chicago: Surrey Books / Agate Publishing, 2023).

p. 64 *"For too long scientists have denied the existence"*: This quotation and the one about fish crows can be found in Jonathan Balcombe, "Animal Pleasure," in *Encyclopedia of Animal Behavior*, ed. Marc Bekoff (Westport, CT: Greenwood Publishing Group, 2004), 563–65.

p. 65 *"Sentient animals have the capacity to experience pleasure"*: Jonathan Leake, "The Secret Life of Moody Cows," *Sunday Times*, February 27, 2005.

p. 65 *Joy in animals may be obvious to one's eyes*: Studies support the idea that play is fun. Neuroscientist Steve Siviy has shown that dopamine (and perhaps serotonin and norepinephrine) is important in the regulation of play; see Steve Siviy, "Neurobiological Substrates of Play Behavior: Glimpses into the Structure and Function of Mammalian Playfulness," in *Animal Play: Evolutionary, Comparative, and Ecological Perspectives*, ed. Marc Bekoff and J. A. Byers (New York: Cambridge University Press, 1998), 221–42.

p. 65 *there's what researcher Patricia Simonet called "a breathy"*: P. Gorner, "Animals Enjoy Good Laugh Too, Scientists Say," *Chicago Tribune*, April 1, 2005; and "Sounds of Dog's 'Laugh'

Endnotes

p. 66   Calms Other Pooches," *ABC News*, December 4, 2005, http://www.abcnews.go.com/GMA/Health/story?id=1370911.

p. 66   *Primatologist Takahisa Matsusaka has shown that play panting*: Takahisa Matsusaka, "When Does Play Panting Occur During Social Play in Wild Chimpanzees?" *Primates* 45 (2004): 221–29.

p. 66   *"Research on rough-housing play in mammals"*: Jaak Panksepp, "Beyond a Joke: From Animal Laughter to Human Joy," *Science* 308 (2005): 62–63.

p. 66   *"It's like the behavior of young children"*: Robert Provine, *Laughter: A Scientific Investigation* (New York: Penguin Books, 2001).

p. 68   *Whether or not other animals display pride is still hotly debated*: Carol Gigliotti, *The Creative Lives of Animals* (New York: New York University Press, 2022).

p. 69   *"Chimpanzees show each other objects just for the sake"*: Nicola Davis, "Cool Leaf! Study Records Chimp Showing Off Object in Human-Like Way," *The Guardian*, November 14, 2022, https://www.theguardian.com/science/2022/nov/14/cool-leaf-study-records-chimp-showing-off-object-in-human-like-way; see also Marc Bekoff, "Do Animals Show Off?," *Psychology Today*, November 16, 2022, https://www.psychologytoday.com/us/blog/animal-emotions/202211/do-animals-show; and Claudia Wilke et al., "Declarative Referential Gesturing in a Wild Chimpanzee (Pan troglodytes)," *Proceedings of the National Academy of Sciences* 119, no. 47 (November 14, 2022), https://www.pnas.org/doi/10.1073/pnas.2206486119.

p. 71   *anecdotal evidence and careful observation indicate feelings akin to awe*: Henry Wismayer, "Finding Awe Amid Everyday Splendor," *Noema*, January 5, 2023, https://www.noemamag.com/finding-awe-amid-everday-splendor.

p. 71   *"As he gets closer, and the roar of the falling water"*: Jane Goodall, "Primate Spirituality," in *The Encyclopedia of Religion and Nature*, ed. B. Taylor (New York: Thoemmes Continuum, 2005), 1303–6.

p. 72   *In 2023, a study was published about rapid spinning*: Adriano Lameira and Marcus Perlman, "Great Apes Reach Momentary Altered Mental States by Spinning," *Primates*, March 14, 2023, https://link.springer.com/article/10.1007/s10329-023-01056-x; and Marc Bekoff, "Great Apes Alter Their Mental States By Spinning Rapidly," *Psychology Today*, March 16, 2023, https://www.psychologytoday.com/us/blog/animal-emotions/202303/great-apes-alter-their-mental-states-by-spinning-rapidly.

p. 72   *"Tragedy struck in 1997. Fourteen's mate, Thirteen"*: Doug Smith, "Meet Five, Nine, and Fourteen, Yellowstone's Heroine Wolves," *Wildlife Conservation* (February 2005): 32.

p. 73   *"It's not anthropomorphic to use this label"*: "Orcas Now Taking Turns Floating Dead Calf in Apparent Mourning Ritual," CBC Radio, July 31, 2018, https://www.cbc.ca/radio/asit happens/as-it-happens-tuesday-edition-1.4768344/orcas-now-taking-turns-floating-dead-calf-in-apparent-mourning-ritual-1.4768349.

p. 73   *Janet Baker-Carr writes in* An Extravagance of Donkeys: Janet Baker-Carr, *An Extravagance of Donkeys* (iUniverse, Inc., 2006).

p. 73   *Some scientists even say that the demeanor of elephants*: C. Siebert, "An Elephant Crackup?," *New York Times Magazine*, October 8, 2006, 42, http://www.nytimes.com/2006/10/08/magazine/08elephant.html; see also G. A. Bradshaw et al., "Elephant Breakdown," *Nature* 433 (2005): 807.

p. 73   *it's known that abandoned dogs can suffer from PTSD*: Nicholas Vincent, "Experts Say Abandoned Dogs Experience Real PTSD," One Green Planet, April 1, 2023, https://www.onegreenplanet.org/animals/experts-say-abandoned-dogs-experience-real-ptsd.

p. 74   *Consider Sadie and Oscar, two senior dogs*: Marc Bekoff, "Do Dogs Know They're Dying?"

Psychology Today, February 14, 2019, https://www.psychologytoday.com/us/blog/animal-emotions/201902/do-dogs-know-theyre-dying.

p. 76　*Researchers from the University of Pennsylvania report that baboons*: A. E. Engh et al., "Behavioural and Hormonal Responses to Predation in Female Chacma Baboons (*Papio hamadryas ursinus*)," *Proceedings of the Royal Society of London: Biological Sciences Series B: Biological Sciences* 273 (2006): 707–12; see also Greg Lester, "Baboons in Mourning Seek Comfort among Friends," University of Pennsylvania, January 30, 2006, https://penntoday.upenn.edu/news/baboons-mourning-seek-comfort-among-friends.

p. 78　*"They stood around Tina's carcass, touching it gently"*: This quotation and the following one by Cynthia Moss are from Cynthia Moss, *Elephant Memories: Thirteen Years in the Life of an Elephant Family* (Chicago: University of Chicago Press, 2000).

p. 78　*Iain Douglas-Hamilton and his colleagues have shown that elephants*: Iain Douglas-Hamilton et al., "Behavioural Reactions of Elephants Towards a Dying and Deceased Matriarch," *Applied Animal Behaviour Science* 100 (2006): 87–102.

p. 78　*Karen McComb, an expert on animal communication and cognition*: Karen McComb, L. Baker, and C. Moss, "African Elephants Show High Levels of Interest in the Skulls and Ivory of Their Own Species," *Biology Letters* (The Royal Society) 2 (2006): 26–28; see also John Pickrell, "Elephants Show Special Interest in Their Dead," *National Geographic News*, October 31, 2005.

p. 79　*In 2018, a mother orca, twenty-year-old Tahlequah*: The details and quotes for this story are from Marc Bekoff, "Make No Mistake, Orca Mom J-35 and Pod Mates Are Grieving," *Psychology Today*, August 1, 2018, https://www.psychologytoday.com/us/blog/animal-emotions/201808/make-no-mistake-orca-mom-j-35-and-pod-mates-are-grieving; Daniel Kreizberg, *Tahlequah the Whale: A Dance of Grief* (Vilnius, Lithuania: Meinhart Animation Studio, 2023), https://tahlequahthewhale.com; Colin Dwyer, "After Calf's Death, Orca Mother Carries It for Days in 'Tragic Tour of Grief,'" *NPR*, July 21, 2018, https://www.npr.org/2018/07/31/634314741/after-calfs-death-orca-mother-carries-it-for-days-in-tragic-tour-of-grief.

p. 79　*Speaking about whales, in December 2022, Port Townsend*: Katie Surma, "Port Townsend Recognizes Rights of Endangered Southern Resident Orcas," *Seattle Times*, December 7, 2022, https://www.seattletimes.com/seattle-news/environment/port-townsend-recognizes-rights-of-endangered-southern-resident-orcas.

p. 81　*One reason we know is because the powerful "love hormone"*: Mi Dog Guide, "What Hormone Is Responsible for the Dog's Loving Behavior?," accessed September 27, 2023, https://midogguide.com/diseases/what-hormone-is-responsible-for-the-dogs-loving-behavior.html.

p. 81　*They don't. In fact, dogs are quite selective*: Marc Bekoff, "Do Pets Really Unconditionally Love and Unwind Us?," *Psychology Today*, June 15, 2020, https://www.psychologytoday.com/us/blog/animal-emotions/202006/do-pets-really-unconditionally-love-and-unwind-us.

p. 82　*While individuals of some species do have love affairs*: Charlyn Partridge, "Sneak Copulation as an Alternative Mating Strategy," in *Encyclopedia of Evolutionary Psychological Science*, ed. T. Shackelford and V. Weekes-Shackelford (Switzerland: Springer, Cham, 2017), https://link.springer.com/referenceworkentry/10.1007/978-3-319-16999-6_3610-1#citeas.

p. 82　*Male marmosets spend a good deal of time figuring out*: Edna Francisco, "Monkey Love: Male Marmosets Think Highly of Sex," *Science News*, February 19, 2004, https://www.sciencenews.org/article/monkey-love-male-marmosets-think-highly-sex; see also C. F. Ferris, et al., "Activation of Neural Pathways Associated with Sexual Arousal in Nonhuman Primates," *Journal of Magnetic Resonance Imaging* 19 (2004): 168–75.

Endnotes

p. 82   *For example, zoologist Bernd Würsig describes the courtship*: Bernd Würsig, "Leviathan Lust and Love," in *The Smile of a Dolphin: Remarkable Accounts of Animal Emotions*, ed. Marc Bekoff (New York: Random House / Discovery Books, 2000), 62–65.

p. 82   *Researcher Lee Dugatkin observed what he calls "guppy love"*: Lee Dugatkin, "Risking It All for Love," in *The Smile of a Dolphin: Remarkable Accounts of Animal Emotions*, ed. Marc Bekoff (New York: Random House / Discovery Books, 2000), 66–67.

p. 82   *"All these data lead me to believe that animals big"*: Helen Fisher, *Why We Love: The Nature and Chemistry of Romantic Love* (New York: Henry Holt, 2004), 47.

p. 83   *Australian leeches, or bloodsuckers, have been shown to be devoted*: "Slimy Leeches Are Devoted Parents," *Daily Times* (Pakistan), March 7, 2004.

p. 83   *One day Naomi saw a mother surface without her calf*: Naomi Rose, "Giving a Little Latitude," in *The Smile of a Dolphin: Remarkable Accounts of Animal Emotions*, ed. Marc Bekoff (New York: Random House / Discovery Books, 2000), 32.

p. 83   *In late February of one year, Echo, the "beautiful matriarch"*: Cynthia Moss, "A Passionate Devotion," in *The Smile of a Dolphin: Remarkable Accounts of Animal Emotions*, ed. Marc Bekoff (New York: Random House / Discovery Books, 2000), 134–37.

p. 85   *A one-year-old hippopotamus, named Owen by his caretakers*: C. Hatkoff, *Owen & Mzee: The Story of a Remarkable Friendship* (New York: Scholastic Press, 2006); see also "Odd Couple Make Friends in Kenya," *BBC News*, January 6, 2005, http://news.bbc.co.uk/2/hi/africa/4152447.stm.

p. 85   *Tika and her longtime mate, Kobuk, two wonderful malamutes*: Anne Bekoff, "In Sickness and in Health," in *The Smile of a Dolphin: Remarkable Accounts of Animal Emotions*, ed. Marc Bekoff (New York: Random House / Discovery Books, 2000), 61–62.

p. 86   *Fifi was a female chimpanzee whom Jane knew*: Jane Goodall, "Pride Goeth Before a Fall," in *The Smile of a Dolphin: Remarkable Accounts of Animal Emotions*, ed. Marc Bekoff (New York: Random House / Discovery Books, 2000), 166–67.

p. 86   *Marc Hauser observed what could be called embarrassment*: Marc Hauser, "If Monkeys Could Blush," in *The Smile of a Dolphin: Remarkable Accounts of Animal Emotions*, ed. Marc Bekoff (New York: Random House / Discovery Books, 2000), 200–1.

p. 87   *bird expert Gisela Kaplan writes about the behavior*: Gisela Kaplan, *Bird Minds: Cognition and Behaviour of Australian Native Birds* (Clayton, Australia: CSIRO Publishing, 2015), 188–89.

p. 87   *Birds are also well known for their sibling conflicts*: For information on sibling conflict, see Douglas Mock, *More Than Kin and Less Than Kind: The Evolution of Family Conflict* (Cambridge, MA: Harvard University Press, 2004).

p. 88   *In spring 2023, orcas made international news for trying to sink boats*: Details for this story are from Hilary Hanson, "Orcas Sink Another Boat in Europe, and the Behavior Is Spreading," *Huffpost*, May 20, 2023, https://www.huffpost.com/entry/orcas-sink-boats-europe_n_6468dc44e4b0ab2b97e7c8a0#; Ruth Estaban et al., "Killer Whales of the Strait of Gibraltar, an Endangered Subpopulation Showing a Disruptive Behavior," *Marine Mammal Science*, June 8, 2022, https://onlinelibrary.wiley.com/doi/10.1111/mms.12947; Kieran Mulvaney, "Vengeance — or Playtime? Why Orcas Are Coordinating Attacks Against Sailboats," *National Geographic*, May 26, 2023, https://www.nationalgeographic.com/animals/article/orcas-killer-whales-boat-attacks; and Marc Bekoff, "Why Are Killer Whales in Spain Bothering Boats?," *Psychology Today*, July 21, 2023, https://www.psychologytoday.com/us/blog/animal-emotions/202307/interview-spanish-ethologist-david-nieto-macein.

p. 89   *Or so says world-famous Stanford professor Robert Sapolsky*: Robert Sapolsky, *A Primate's Memoir* (New York: Touchstone Books, 2002), 234.

p. 90   *For instance, in Saudi Arabia, a man hit and killed a baboon*: "Revenge Attack by Stone-Throwing Baboons," *Ananova*, December 9, 2000.

p. 90   *Ron Schusterman, a marine biologist at the University of California*: Ron Schusterman, "Pitching a Fit," in *The Smile of a Dolphin: Remarkable Accounts of Animal Emotions*, ed. Marc Bekoff (New York: Random House / Discovery Books, 2000), 106–7.

p. 91   *Simon Baron-Cohen has made great strides*: Simon Baron-Cohen, *Mindblindness: An Essay on Autism and Theory of Mind* (Cambridge, MA: MIT Press, 1995); and E. A. Tinbergen and N. Tinbergen, "Early Childhood Autism — An Ethological Approach," *Advances in Ethology* 10 (1972).

Chapter 4: Wild Justice, Empathy, and Fair Play

Some of the material in this chapter relies on my books *Minding Animals*, *Animal Passions and Beastly Virtues*, and *Wild Justice*, with Jessica Pierce, all of which are listed in the bibliography.

p. 93   *"Those communities which included the greatest number"*: Charles Darwin, *The Descent of Man and Selection in Relation to Sex* (New York: Random House, 1871/1936), 163.

p. 94   *Other ethologists (such as... the well-known field biologist George Schaller)*: George Schaller and G. R. Lowther, "The Relevance of Carnivore Behavior to the Study of Early Hominids," *Southwestern Journal of Anthropology* 25 (1969): 307–41.

p. 97   *For instance, an orangutan named Fu Manchu*: Delana Lefevers, "Fu Manchu, the Famed Orangutan Escape Artist, Learned to Pick Locks Right Here in Nebraska," Only In Your State, March 8, 2023, https://www.onlyinyourstate.com/nebraska/fu-manchu-orangutan-ne.

p. 97   *Wild gorillas, similarly, have been seen dismantling poachers' traps*: Ker Than, "Gorilla Youngsters Seen Dismantling Poachers' Traps — A First," *National Geographic*, July 18, 2012, https://www.nationalgeographic.com/animals/article/120719-young-gorillas-juvenile-traps-snares-rwanda-science-fossey.

p. 101   *In fact, in 2015, an entire issue of the scientific journal* Current Biology: "Biology of Fun," *Current Biology* 25, no. 1 (January 5, 2015), https://www.sciencedirect.com/journal/current-biology/vol/25/issue/1.

p. 101   *"If it looks like fun and has any other necessary attributes of fun"*: David Matheson, "If It Looks Like Animal Fun, Then It Probably Is Fun" (letter to the editor), *New Scientist*, no. 3424 (February 4, 2023), https://www.newscientist.com/letter/mg25734241-100-if-it-looks-like-animal-fun-then-it-probably-is-fun.

p. 101   *In her book* On Talking Terms with Dogs, *Norwegian dog trainer*: Turid Rugaas, *On Talking Terms with Dogs: Calming Signals* (Wenatchee, WA: Dogwise Publishing, 2005).

p. 101   *Renowned neurobiologist Jaak Panksepp discovered a close association*: For more, see Jaak Panksepp, *Affective Neuroscience* (New York: Oxford University Press, 1998); and Jaak Panksepp, "Beyond a Joke: From Animal Laughter to Human Joy," *Science* 308 (2005): 62–63.

p. 102   *James Rilling and his colleagues have used functional magnetic*: James Rilling et al., "A Neural Basis for Cooperation," *Neuron* 36 (2002): 395–405.

p. 102   *Also, researchers have identified a "trust center"*: B. D. King-Casas et al., "Getting to Know You: Reputation and Trust in a Two-Person Economic Exchange," *Science* 308 (2005): 78;

Endnotes

see also Matthew Herper, "Measuring Trust with a Brain Scan," *Forbes*, March 31, 2005, https://www.forbes.com/2005/03/31/cx_mh_0331brain.html.

p. 103 *Bows serve as punctuation, an exclamation point*: Marc Bekoff, "Play Signals as Punctuation: The Structure of Social Play in Canids," *Behaviour* 132 (1995): 419–29.

p. 103 *In a 2016 study of pairs of adult dogs, Sarah-Elizabeth Byosiere*: Sarah-Elizabeth Byosiere, Julia Espinosa, and Barbara Smuts, "Investigating the Function of Play Bows in Adult Pet Dogs (*Canis lupus familiaris*)," *Behavioural Processes* 125 (April 2016): 106–13, https://pubmed.ncbi.nlm.nih.gov/26923096.

p. 103 *Jessica Flack and her colleagues discovered that juvenile chimpanzees*: Jessica Flack, L. A. Jeannotte, and Frans de Waal, "Play Signaling and the Perception of Social Rules by Juvenile Chimpanzees (*Pan troglodytes*)," *Journal of Comparative Psychology* 118 (2004): 149–59.

p. 104 *Red-necked wallabies, kangaroos of a kind, engage*: Duncan Watson and D. B. Croft, "Age-Related Differences in Playfighting Strategies of Captive Male Red-Necked Wallabies (*Macropus rufogriseus banksianus*)," *Ethology* 102 (1996): 336–46.

p. 105 *Psychologist Sergio Pellis discovered that sequences of rat play*: Sergio Pellis, "Keeping in Touch: Play Fighting and Social Knowledge," in *The Cognitive Animal: Empirical and Theoretical Perspectives on Animal Cognition*, ed. Marc Bekoff, Colin Allen, and Gordon M. Burghardt (Cambridge, MA: MIT Press, 2002), 421–27.

p. 105 *Along these lines, two colleagues and I proposed that play*: Marek Spinka, Ruth C. Newberry, and Marc Bekoff, "Mammalian Play: Training for the Unexpected," *Quarterly Review of Biology* 76, no. 2 (June 2001): 141–68, https://pubmed.ncbi.nlm.nih.gov/11409050.

p. 106 *Research on nonhuman primates has shown that punishment*: For more on this research, see the following: R. Sussman and P. A. Garber, "Rethinking Sociality: Cooperation and Aggression Among Primates," in *The Origins and Nature of Sociality*, ed. R. Sussman and A. Chapman (Chicago: Aldine, 2004), 161–91; and R. Sussman, P. A. Garber, and J. Cheverud, "The Importance of Cooperation and Affiliation in the Evolution of Primate Sociality," *American Journal of Physical Anthropology* 128 (2005): 84–97.

p. 107 *Alexandra Horowitz observed a dog she called Up-ears*: Alexandra Horowitz, "The Behaviors of Theories of Mind, and a Case Study of Dogs at Play" (PhD diss., University of California, San Diego, 2002).

p. 107 *Among dogs, the moral landscape of play is Don't bow*: Marc Bekoff, "When Dogs Play, They Follow the Golden Rules of Fairness," *Psychology Today*, November 15, 2019, https://www.psychologytoday.com/us/blog/animal-emotions/201911/when-dogs-play-they-follow-the-golden-rules-fairness.

p. 109 *For example, my long-term field research on coyotes*: This is called the "social cohesion" theory of dispersal; see Marc Bekoff, "Mammalian Dispersal and the Ontogeny of Individual Behavioral Phenotypes," *American Naturalist* 111 (1997): 715–32.

p. 111 *"I believe that at the most fundamental level"*: His Holiness the Dalai Lama, "Understanding Our Fundamental Nature," in *Visions of Compassion: Western Scientists and Tibetan Buddhists Examine Human Nature*, ed. R. J. Davidson and A. Harrington (New York: Oxford University Press, 2002), 68.

p. 111 *"We believe that most donkeys"*: Michael Tobias and Jane Morrison, *Donkey: The Mystique of* Equus asinus (Tulsa, OK: Council Oak Books, 2006), 42.

p. 111 *Ernst Fehr and his colleagues have discovered that when treated fairly*: Ernst Fehr and S. Gächter, "Altruistic Punishment in Humans," *Nature* 415 (2002): 137–40; Ernst Fehr and B. Rockenbach, "Detrimental Effect of Sanctions on Human Altruism," *Nature* 422 (2003):

p. 112   137–40; and K. Sigmund, E. Fehr, and M. A. Nowak, "The Economics of Fair Play," *Scientific American* 286 (2002): 83–87.

p. 112   *Felix Warneken and Michael Tomasello discovered that infants*: Felix Warneken and Michael Tomasello, "Altruistic Helping in Human Infants and Young Chimpanzees," *Science* 311 (2006): 1301.

p. 112   *Primatologists Robert Sussman and Paul Garber report that for diurnal*: Robert Sussman and Paul Garber, "Rethinking Sociality: Cooperation and Aggression Among Primates," in *The Origins and Nature of Sociality*, ed. R. Sussman and A. Chapman (Chicago: Aldine, 2004), 161–91; and Robert Sussman, Paul Garber, and J. Cheverud, "The Importance of Cooperation and Affiliation in the Evolution of Primate Sociality," *American Journal of Physical Anthropology* 128 (2005): 84–97.

p. 112   *In their landmark 2020 book* Survival of the Friendliest: Marc Bekoff, "Friendliness and Cooperation Are Secrets of Humans' Success," *Psychology Today*, July 13, 2020, https://www.psychologytoday.com/us/blog/animal-emotions/202007/friendliness-and-cooperation-are-secrets-humans-success; Marc Bekoff, "Trust and Cooperation Are Widespread Among Diverse Animals," *Psychology Today*, September 21, 2022, https://www.psychologytoday.com/us/blog/animal-emotions/202209/trust-and-cooperation-are-widespread-among-diverse-animals; see also Kristin Ohlson, *Sweet in Tooth and Claw: Stories of Generosity and Cooperation in the Natural World* (Manchester, VT: Patagonia, 2022); and Nichola Raihani, *The Social Instinct: How Cooperation Shaped the Word* (New York: St. Martin's Press, 2021).

p. 113   *Some ecologists take this even further*: B. Shouse, "Ecology: Conflict Over Cooperation," *Science* 299 (2003): 644–46.

p. 114   *"Justice presumes a personal concern for others"*: Robert Solomon, *A Passion for Justice: Emotions and the Origins of the Social Contract* (Lanham, MD: Rowman & Littlefield Publishers, 1995), 102.

p. 114   *"Forgiveness has a biological foundation"*: David Sloan Wilson, *Darwin's Cathedral: Evolution, Religion, and the Nature of Society* (Chicago: University of Chicago Press, 2002), 195, 212.

Chapter 5: Addressing Uncertainty and Anthropomorphism in the Study of Animal Minds

p. 117   *"Sometimes I read about someone saying with great authority"*: Claudia Dreifus, "A Conversation with Frans de Waal," *New York Times*, June 26, 2001, https://www.nytimes.com/2001/06/26/science/a-conversation-with-frans-de-waal-observing-the-behavior-of-apes-up-close.html.

p. 125   *"It is possible, therefore, that your simple man"*: William J. Long, *Brier-Patch Philosophy by "Peter Rabbit"* (Boston and London: Ginn & Company, 1906), 26.

p. 125   *"In the late 1990s two remarkable novels were published"*: Hal Whitehead, *Sperm Whales: Social Evolution in the Oceans* (Chicago: University of Chicago Press, 2004), 370–71.

p. 126   *Lorenz noted that human beings are attracted*: Konrad Lorenz, "Ganzheit und Teil in der tierischen und menschlichen Gemeinschaft," 1950. Reprinted in *Studies in Animal and Human Behaviour*, vol. 2, ed. R. Martin (Cambridge, MA: Harvard University Press, 1971), 135.

p. 127   *"Do I get grief for the fact that in communicating"*: Douglas Cruickshank, "Robert Sapolsky," *Salon*, May 14, 2001, https://www.salon.com/2001/05/14/sapolsky; see also Robert Sapolsky, *A Primate's Memoir* (New York: Touchstone Books, 2002).

p. 127   *"Yes, we are human and cannot avoid the language"*: Stephen J. Gould, "A Lover's Quarrel," in *The Smile of a Dolphin: Remarkable Accounts of Animal Emotions*, ed. Marc Bekoff (New York: Random House / Discovery Books, 2000), 13–17.

p. 127 *Anthropomorphism endures because it is a necessity*: In her book *Why Animals Matter* (Oxford, UK: Oxford University Press, 2012), Marian Dawkins falsely claims that "others, most notably Marc Bekoff, go in for full-blooded, genuine anthropomorphism" (p. 21). Dawkins goes on to write that my and others' brand of anthropomorphism "may well be right," but then mischaracterizes my position as being "virtually tantamount to saying that there are no limits to how we interpret animal behaviour" (p. 22). She also says I've set up an "anthropomorphism vs. science" dichotomy. Nothing is further from the truth. I've always maintained that using anthropomorphic language allows the behavior and emotions of other animals to be accessible to us, and I've always stressed that we can be biocentrically anthropomorphic and do rigorous science. For my full response to Dawkins, see Marc Bekoff, "Animal Consciousness and Science Matter," *Psychology Today*, May 7, 2012, https://www.psychologytoday.com/us/blog/animal-emotions/201205/animal-consciousness-and-science-matter.

p. 128 *"I have learned that anthropomorphism is a deeply suspect word"*: Marc Bekoff and Koen Margodt, eds., *Jane Goodall at 90: Celebrating an Astonishing Lifetime of Science, Advocacy, Humanitarianism, Hope, and Peace* (Berlin, MD: Salt Water Media, 2024).

p. 128 *When it comes to anthropomorphic language*: Marc Bekoff, "Anthropomorphic Double-Talk: Can Animals Be Happy but Not Unhappy? No!," *Psychology Today*, June 24, 2009, https://www.psychologytoday.com/us/blog/animal-emotions/200906/anthropomorphic-double-talk-can-animals-be-happy-not-unhappy-no.

p. 128 *A good example is the story of Ruby, a forty-three-year-old African elephant*: Patricia Ward Biederman, "Soft Heart Under Her Thick Skin?" *Los Angeles Times*, November 16, 2004, https://www.latimes.com/archives/la-xpm-2004-nov-16-me-ruby16-story.html.

p. 129 *"Obviously the mahouts [elephant keepers] may have many beliefs"*: Mary Midgley was quoted by Animal Sentience on its website on September 4, 2005.

p. 130 *I'm happy to report that because of the activism to free her*: Lisa Brenner, "Ruby the Elephant, 50, Dies at Animal Sanctuary," *LAist*, March 31, 2011, https://laist.com/news/ruby-the-elephant-50-dies-at-animal.

p. 130 *Psychologist Gordon Burghardt notes that denying our own intuitions*: Gordon Burghardt, "Animal Awareness: Current Perceptions and Historical Perspective," *American Psychologist* 40 (1985): 905–19.

p. 130 *Burghardt calls for "critical anthropomorphism"*: Gordon Burghardt, "Mediating Claims through Critical Anthropomorphism," *Animal Sentience* 3, no. 17 (2016), https://www.wellbeingintlstudiesrepository.org/cgi/viewcontent.cgi?article=1063&context=animsent; and Marc Bekoff, "Animal Emotions: Exploring Passionate Natures," *BioScience* 50, no. 10 (October 2000), https://academic.oup.com/bioscience/article/50/10/861/233998.

p. 130 *The renowned and influential experimental psychologist Donald Hebb*: Donald Hebb, "Emotion in Man and Animal: An Analysis of the Intuitive Process of Recognition," *Psychological Review* 53 (1946): 88–106.

p. 131 *A very interesting and important 2019 study*: Marc Bekoff, "Anthropomorphism Favors Coexistence, Not Deadly Domination," *Psychology Today*, December 28, 2019, https://www.psychologytoday.com/us/blog/animal-emotions/201912/anthropomorphism-favors-cexistence-not-deadly-domination.

p. 131 *In their research essay published in* Biological Conservation: Brandon Keim, "America's Views on Wildlife Are Changing," *Anthropocene*, December 18, 2019, https://www.anthropocenemagazine.org/2019/12/anthropomorphism-and-wildlife.

p. 132 *"Anthropomorphism has the potential to aid conservation"*: Alvin A. Y.-H. Chan, "Anthropomorphism as a Conservation Tool," *Nature Reviews* 21, no. 7 (June 2012).

p. 132 *"Animals embody every quality found in the human personality"*: Paul Shepard is quoted in Cody Moser, "On the Origins of Empathy for Other Species," *Works in Progress* 10, February 24, 2023, https://worksinprogress.co/issue/on-the-origins-of-empathy-for-other-species.

p. 132 *"We know that nonhuman animals lead rich emotional lives"*: Marlon H. Reis, "The Importance of Recognizing Sentience in Non-Human Animals," *Marc Bekoff, Ph.D.* (blog), March 8, 2023, https://marcbekoff.com/marcs-essays/f/the-importance-of-recognizing-sentience-in-non-human-animals.

Chapter 6: Ethical Choices

p. 133 *"Ethics in our Western world has hitherto"*: Albert Schweitzer, *Memoirs of Childhood and Youth* (London: Allen and Unwin, 1924).

p. 134 *It was only a hundred years ago that writers and naturalists*: For more on the "nature faker" movement, see R. H. Lutts, *The Nature Fakers* (Golden, CO: Fulcrum Publishers, 1990); and R. K. Gould, *At Home in Nature: Modern Homesteading and Spiritual Practice in America* (Berkeley: University of California Press, 2005).

p. 135 *a "wounded world," as the ecologist Paul Ehrlich*: Paul Ehrlich, *A World of Wounds: Ecologists and the Human Dilemma* (Oldendorf/Luhe, Germany: Ecology Institute, 1997).

p. 135 *"The nature of science is that it never"*: John Webster, "Animal Sentience and Animal Welfare: What Is It to Them and What Is It to Us?," *Applied Animal Behaviour Science* 100 (2006): 1–3; see also John Webster, *Animal Welfare: Limping Towards Eden* (Oxford, England: Blackwell Publishing, 2005).

p. 135 *"Because uncertainty never disappears, decisions"*: Henry Pollack, *Uncertain Science... Uncertain World* (New York: Cambridge University Press, 2003), 3.

p. 138 *In March 2016, China released its first set of guidelines*: Kathleen McLaughlin, "China Finally Setting Guidelines for Treating Lab Animals," *Science*, March 21, 2016, https://www.science.org/content/article/china-finally-setting-guidelines-treating-lab-animals; Amanda Lindner, "Oh Yeah! Iran Bans Use of Wild Animals in Circuses!," One Green Planet, accessed September 28, 2023, https://www.onegreenplanet.org/news/iran-bans-use-of-wild-animals-in-circuses.

p. 138 *"the assessment of welfare can be carried out in an objective"*: Donald Broom, *Broom and Fraser's Domestic Animal Behaviour and Welfare*, sixth edition (Oxfordshire, UK: CABI, 2007/2022), 6.

p. 138 *In 1965, the British Farm Animal Welfare Council created*: Melissa Elischer and Tina Conklin, "The Five Freedoms: A History Lesson in Animal Care and Welfare," Michigan State University Extension, September 6, 2019, https://www.canr.msu.edu/news/an_animal_welfare_history_lesson_on_the_five_freedoms.

p. 139 *In* Unleashing Your Dog, *Jessica and I expand the Five Freedoms*: Marc Bekoff, "Dogs, Captivity, and Freedom: Unleash Them Whenever You Can," *Psychology Today*, March 15, 2019, https://www.psychologytoday.com/us/blog/animal-emotions/201903/dogs-captivity-and-freedom-unleash-them-whenever-you-can.

p. 139 *These ten freedoms are echoed by Martha C. Nussbaum's "capabilities"*: Marc Bekoff, "Justice for Animals Means Freedom to Do What Comes Naturally," *Psychology Today*, December 12, 2022, https://www.psychologytoday.com/us/blog/animal-emotions/202212/justice-animals-means-freedom-do-what-comes-naturally.

Endnotes

p. 140 *In his landmark book* Respect for Nature, *philosopher Paul Taylor*: All the following quotes by Paul Taylor are from Paul Taylor, *Respect for Nature: A Theory of Environmental Ethics* (Princeton, NJ: Princeton University Press, 1986/2011), 173, 174, 175, 186.

p. 140 *For example, Jane Goodall claims that her relationship*: Jane Goodall, "Digging Up the Roots," *Orion* 13 (Winter 1994): 20–21.

p. 141 *"Certainly it seems like a dirty double-cross"*: Lawrence Johnson, *A Morally Deep World* (New York: Cambridge University Press, 1991), 122.

p. 141 *Around 100 million animals are used in laboratory research globally*: It is very difficult to get accurate counts of animals used in labs. The statistics in this paragraph combine details from several sources (all accessed on September 28, 2023): World Day for Laboratory Animals 2023, https://www.congressweb.com/animaldefendersinternational/5; "Animals in Laboratories," Animal Welfare Institute, https://awionline.org/content/animals-laboratories; Cruelty Free International, "The US Is Still Failing Animals and the Public," February 24, 2023, https://crueltyfreeinternational.org/latest-news-and-updates/us-still-failing-animals-and-public; and Larry Carbone, "Estimating Mouse and Rat Use in American Laboratories by Extrapolation from Animal Welfare Act–Regulated Species," *Scientific Reports* 11, no. 493 (2021), https://www.nature.com/articles/s41598-020-79961-0.

p. 142 *"We are amending the Animal Welfare Act (AWA) regulations"*: Department of Agriculture, "Animal Welfare, Definition of Animal," RIN0579-AB69, *Federal Register*, vol. 69, no. 108, June 4, 2004, https://www.govinfo.gov/content/pkg/FR-2004-06-04/html/04-12693.htm.

p. 142 *For instance, the Animal Welfare Act legally allows chimpanzees*: For the complete text of the Animal Welfare Act, visit U.S. Department of Agriculture at http://www.nal.usda.gov/awic/legislat/usdaleg1.htm.

p. 142 *In response to the discovery that mice are empathic rodents*: Ishani Ganguli, "Mice Show Evidence of Empathy," *The Scientist*, June 30, 2006; and D. J. Langford et al., "Social Modulation of Pain as Evidence for Empathy in Mice," *Science* 312 (2006): 1967–70.

p. 143 *"The observed changes of acoustical parameters"*: B. Puppe et al., "Castration-Induced Vocalisation in Domestic Piglets (*Sus scrofa*): Complex and Specific Alterations of the Vocal Quality," *Applied Animal Behaviour Science* 95 (2005): 67–78.

p. 143 *For instance, using a $500,000 taxpayer-funded grant*: Associated Press, "Tasers to Be Tested on Cocaine-Drugged Pigs," *NBC News*, May 3, 2005, https://www.nbcnews.com/id/wbna7405567.

p. 143 *Then there were psychologist Harry Harlow's well-known maternal*: For a summary of Harry Harlow's research, see D. Blum, *Love at Goon Park: Harry Harlow and the Science of Affection* (New York: Perseus Publishing, 2002).

p. 144 *In February 2006, the prestigious Diabetes Research Institute*: Diabetes Research Institute, "Researchers Find Striking Difference between Human and Animal Insulin-Producing Islet Cells," February 2006.

p. 144 *And in an essay published in the* Journal of the American Medical: J. Lazarou, B. H. Pomeranz, and P. N. Corey, "Incidence of Adverse Drug Reactions in Hospitalized Patients," *Journal of the American Medical Association* 279 (1998): 1205.

p. 144 *In 2023, it was reported that the unnatural and sterile*: Sonia Shah, "The Case for Free-Range Lab Mice," *New Yorker*, February 18, 2023, https://www.newyorker.com/culture/annals-of-inquiry/the-case-for-free-range-lab-mice.

p. 144 *It turns out male mice, usually the default choice*: Azeen Ghorayshi, "Guess Which Sex Behaves More Erratically (at Least in Mice)," *New York Times*, March 7, 2023, https://www.nytimes.com/2023/03/07/science/female-mice-hormones.html.

p. 144   *In a 2023 essay called "Could the Next Blockbuster Drug"*: Emily Anthes, "Could the Next Blockbuster Drug Be Lab-Rat Free?," *New York Times*, March 7, 2023, https://www.nytimes.com/2023/03/07/health/drug-animals-testing.html.

p. 144   *For years, there were unsubstantiated allegations*: The details of this story are based on personal communications between Matt Rossell and the author.

p. 146   *However, I agree with neurologist Dr. Aysha Akhtar*: Brandon Kiem, "What Do We Owe Lab Animals?," *New York Times*, January 23, 2023, https://www.nytimes.com/2023/01/23/science/what-do-we-owe-lab-animals.html.

p. 146   *"Sound science requires animal subjects to be physically"*: Kate Field et al., "Publication Reform to Safeguard Wildlife from Researcher Harm," *PLOS Biology* (April 11, 2019), https://journals.plos.org/plosbiology/article?id=10.1371/journal.pbio.3000193.

p. 147   *But fieldwork, like research on captive animals, isn't always user-friendly*: Benjamin G. Farrar, Ljerka Ostojić, and Nicola S. Clayton, "The Hidden Side of Animal Cognition Research: Scientists' Attitudes Toward Bias, Replicability, and Scientific Practice," *PLOS ONE* (August 31, 2021), https://journals.plos.org/plosone/article?id=10.1371/journal.pone.0256607.

p. 147   *It seems silly to have to remind people that wild animals*: Andrea Harvey et al., "Mental Experiences in Wild Animals: Scientifically Validating Measurable Welfare Indicators in Free-Roaming Horses," *Animals* 13, no. 9 (April 2023), https://www.mdpi.com/2076-2615/13/9/1507; and University of Technology Sydney, "Assessing Emotions in Wild Animals," *Science Daily*, May 17, 2023, https://www.sciencedaily.com/releases/2023/05/230517122134.htm.

p. 147   *It's also important to point out that when researchers negatively impact*: David Foster, "Wildlife Researchers' Good Intentions Often Can Be Deadly for Animals," *Los Angeles Times*, July 7, 2002, https://www.latimes.com/archives/la-xpm-2002-jul-07-adme-studied7-story.html. This piece has stood the test of time and offers plenty of good examples of how research methods influence the behavior of the animals being studied, including how wing tags induced ruddy ducks to spend less time courting and more time sleeping; ear tags on white-footed mice kept them from grooming away ticks; mallards wearing one-ounce radio transmitters rested and preened more, started their nests later, and laid fewer and smaller eggs; and scientists studying endangered Hawaiian honeycreepers used radio transmitters with long, limp antennas, only to find birds hanging helplessly from tree limbs, entangled by the antennas.

p. 147   *Picking up a mouse by their tail, something numerous researchers*: Alla Katsnelson, "Science Is Just Starting to Understand What Animals Feel," *Science News Explores*, October 6, 2022, https://www.snexplores.org/article/animal-emotion-behavior-welfare-feelings.

p. 147   *"Worse still: human-induced fear can lead to natural"*: Gisela Kaplan, "Fieldwork Results, Anonymity, Rare Observations and Cognition-Questions of Method, Biases and Interpretations," In&Sight, June 29, 2022, https://inandsight.science/journal/papers/62b9a4b5028eec53f7fa276f.

p. 149   *Animals value their freedom like we do, and they suffer*: Some of this material is taken from Jessica Pierce and Marc Bekoff, "Incarcerating Animals and Egregious Losses of Freedoms," in *Carceral Logics: Human Incarceration and Animal Captivity*, ed. Lori Gruen and Justin Marceau (New York: Cambridge University Press, 2022), https://www.researchgate.net/publication/359958734_Incarcerating_Animals_and_Egregious_Losses_of_Freedoms.

p. 149   *Scientist Françoise Wemelsfelder notes that researchers often use the neutral*: The quotes here are from Françoise Wemelsfelder, "Animal Boredom: Understanding the Tedium of

Endnotes

Confined Lives," in *Mental Health and Well-Being in Animals*, ed. F. McMillan (Oxford, England: Blackwell Publishing, 2005), 79–93.

p. 150 *In a series of excellent research papers on abnormal repetitive*: Georgia Mason et al., "Why and How Should We Use Environmental Enrichment to Tackle Stereotyped Behaviour?" *Applied Animal Behaviour Science* 102 (2007).

p. 150 *In one Dutch scientific laboratory*: Françoise Wemelsfelder, "Animal Boredom: Understanding the Tedium of Confined Lives," in *Mental Health and Well-Being in Animals*, ed. F. McMillan (Oxford, England: Blackwell Publishing, 2005), 79–93.

p. 150 *Professor Charles Gross conducted a study of marmosets*: Y. Kozorovitskiy et al., "Experience Induces Structural and Biochemical Changes in the Adult Primate Brain," *Proceedings of the National Academy of Sciences* 102 (2005): 17478–82; see also "Bored Monkeys Make for Stupid Monkeys," *New Scientist*, November 16, 2005, http://www.newscientist.com/channel/being-human/brain/mg18825266.100.

p. 151 *Yet any science must provide a description of its domain*: Dale Jamieson and Marc Bekoff, "On Aims and Methods of Cognitive Ethology," *Proceedings of the Biennial Meeting of the Philosophy of Science Association 1992, Volume Two: Symposia and Invited Papers* (1992), 110–24, https://as.nyu.edu/content/dam/nyu-as/faculty/documents/JamiesonBekoff1992.pdf.

p. 152 *James Rilling and his colleagues have used PET scans*: James Rilling, J. T. Winslow, and C. D. Kilts, "The Neural Correlates of Mate Competition in Dominant Male Rhesus Macaques," *Biological Psychiatry* 56 (2004): 363–75.

p. 152 *Also, in one study done on living dolphins*: Dorian Houser et al., "Structural and Functional Imaging of Bottlenose Dolphin (*Tursiops truncatus*) Cranial Anatomy," *Journal of Experimental Biology* 207 (October 2004): 3657–65.

p. 152 *While noninvasive fMRI neuroimaging has been used*: Madalena E. Fonseca et al., "Functional MRI of Large Scale Activity in Behaving Mice," Prelights, April 18, 2020, https://prelights.biologists.com/highlights/functional-mri-of-large-scale-activity-in-behaving-mice; Patricia Pais-Roldán et al., "Contribution of Animal Models toward Understanding Resting State Functional Connectivity," *NeuroImage* 245 (December 15, 2021), https://www.sciencedirect.com/science/article/pii/S1053811921009034; and for different studies on dogs, see Marc Bekoff, "How Dogs View the World: Brain Scans Tell Us What They See," *Psychology Today*, April 14, 2028, https://www.psychologytoday.com/us/blog/animal-emotions/201804/how-dogs-view-the-world-brain-scans-tell-us-what-they-see.

p. 153 *In another study of jealousy, this one on dogs*: Marc Bekoff, "Dogs Know When They've Been Dissed, and Don't Like It a Bit," *Psychology Today*, July 23, 2014, https://www.psychologytoday.com/us/blog/animal-emotions/201407/dogs-know-when-theyve-been-dissed-and-dont-it-bit; Peter Cook et al., "Jealousy in Dogs? Evidence from Brain Imaging," *Animal Sentience* 22, no. 1 (2018), https://www.wellbeingintlstudiesrepository.org/cgi/viewcontent.cgi?article=1319&context=animsent; and Marc Bekoff, "Jealousy in Dogs: Brain Imaging Shows They're Similar to Us," *Psychology Today*, May 13, 2018, https://www.psychologytoday.com/us/blog/animal-emotions/201805/jealousy-in-dogs-brain-imaging-shows-theyre-similar-us.

p. 154 *"If slaughterhouses had glass walls"*: PETA, "'Glass Walls' with Paul McCartney" (video), accessed September 29, 2023, https://www.peta.org/videos/glass-walls-2.

p. 154 *Many solid exposés of factory farming*: Discussions about factory farming can be found in books by Gail Eisner, Michael W. Fox, Peter Singer and Jim Mason, Michael Pollan, Karen Davis, Eric Marcus, Matthew Scully, and Jane Goodall, among others. Animals other than cattle, pigs, and chickens need also to be considered in discussions of sentience.

For example, there is evidence that lobsters and fish feel fear and pain: see T. Corson, *The Secret Life of Lobsters* (New York: Harper Perennial, 2004); S. Yue, R. D. Moccia, and I. J. H. Duncan, "Investigating Fear in Domestic Rainbow Trout (*Oncorhynchus mykiss*), Using an Avoidance Learning Task," *Applied Animal Behaviour Science* 87 (2004): 343–54; and L. U. Sneddon, "The Evidence for Pain in Fish: The Use of Morphine as an Analgesic," *Applied Animal Behaviour Science* 83 (2003): 153–62.

p. 155 *Today, while estimates vary, we are killing anywhere*: The numbers in this section come from several sources: Grace Hussain, "How Many Animals Are Killed for Food Every Day?," *Sentient Media*, August 31, 2022, https://sentientmedia.org/how-many-animals-are-killed-for-food-every-day; Marc Bekoff, "The Effects of Food on Ecosystems and Biodiversity," *Psychology Today*, December 13, 2020, https://www.psychologytoday.com/gb/blog/animal-emotions/202012/the-effects-food-ecosystems-and-biodiversity; and Animalclock.org, which keeps a running toll of animals killed in the United States.

p. 155 *"A team headed by Prof. Ron Milo found that the biomass"*: "The Weight of Responsibility: Biomass of Livestock Dwarfs That of Wild Mammals," American Committee for the Weizmann Institute of Science, February 27, 2023, https://www.weizmann-usa.org/news-media/news-releases/the-weight-of-responsibility-biomass-of-livestock-dwarfs-that-of-wild-mammals.

p. 156 *According to Lucas Reijinders and Sam Soret, compared to soy*: Lucas Reijinders and Sam Soret, "Quantification of the Environmental Impact of Different Dietary Protein Choices," *Journal of Clinical Nutrition* 78 (2003): 664S–68S.

p. 156 *"A single dairy cow belches and farts 114 kilos"*: James Bartholomew, "Let's Ban All the Methane Machines," *Telegraph*, April 7, 2004.

p. 156 *In 2023, it was estimated that the total population*: Oliver Jones, "How Much Methane Do Cows Produce? Scientific Facts," PetKeen, September 19, 2023, https://petkeen.com/how-much-methane-do-cows-produce.

p. 156 *"Food production accounts for over a quarter"*: University of Cambridge, "New Animal Welfare Scoring System Could Enable Better-Informed Food and Farming Choices," *Science News*, March 22, 2023, https://www.sciencedaily.com/releases/2023/03/230322082713.htm.

p. 156 *Since 1970, human activities have wiped out 60 percent*: These statistics are from Laila Kassam and Amir Kassam, "Introduction," in *Rethinking Food and Agriculture*, ed. Amir Kassam and Laila Kassam (Cambridge, MA: Woodhead Publishing / Elsevier, 2020), https://www.sciencedirect.com/science/article/pii/B9780128164105099874. In addition, the majority of water in the Colorado River basin goes to feeding livestock; Elena Shao, "The Colorado River Is Shrinking. See What's Using All the Water," *New York Times*, May 22, 2023, https://www.nytimes.com/interactive/2023/05/22/climate/colorado-river-water.html.

p. 156 *The worst slums in the human world do not compare*: For more information on conditions for poultry, see United Poultry Concerns, http://upc-online.org.

p. 157 *However, if you want a glimpse of what this entails*: Nicholas Kristof, "Spy Cams Show What the Pork Industry Tries to Hide," *New York Times*, February 4, 2023, https://www.nytimes.com/2023/02/04/opinion/meat-pork-animal-farm.html.

p. 157 *in 2023, a new animal welfare scoring system*: University of Cambridge, "New Animal Welfare Scoring System Could Enable Better-Informed Food and Farming Choices," *Science News*, March 22, 2023, https://www.sciencedaily.com/releases/2023/03/230322082713.htm.

p. 157 *Another supposedly positive example is the use of warmer incubators*: Christa Lesté-Lasserre, "Male Chicks Are More Sociable If They Were Grown in Warmer Eggs," *New Scientist*, January 17, 2023, https://www.newscientist.com/article/2355182-male-chicks-are-more-sociable-if-they-were-grown-in-warmer-eggs.

Endnotes

p. 158 — *Some veterinarians have called pigs "gastronomic godsends"*: Mark Essig, *Lesser Beasts* (New York: Basic Books, 2015).

p. 158 — *"Everything we do is about minimising the stress"*: Jane McNaughton, "Pig Slaughter Methods Defended by Pork Industry after Animal Activists Release Footage," *ABC Rural News*, March 27, 2023, https://www.abc.net.au/news/rural/2023-03-28/pig-slaughter-methods-defended-by-pork-industry/102153822.

p. 158 — *In May 2023, the US Supreme Court supported California's*: Jennifer Shike, "BREAKING: Supreme Court Backs California Prop 12," *Farm Journal's Pork*, May 11, 2023, https://www.porkbusiness.com/news/ag-policy/breaking-supreme-court-backs-california-prop-12.

p. 158 — *"I think using animals for food is an ethical thing to do"*: This quote is repeated often without attribution; one source is https://www.goodreads.com/quotes/419241-i-think-using-animals-for-food-is-an-ethical-thing.

p. 159 — *"As we talked of freedom and justice one day"*: Alice Walker, "Am I Blue?," in *Living by the Word* (New York: Harcourt Brace Jovanovich Publishers, 1988), 8.

p. 159 — *I decided to become a vegetarian and then vegan*: Marc Bekoff, "Going 'Cold Tofu' to End Factory Farming," *Psychology Today*, October 31, 2011, https://www.psychologytoday.com/us/blog/animal-emotions/201110/going-cold-tofu-end-factory-farming.

p. 159 — *Many people are also surprised to learn of the horrors*: Marc Bekoff, "The Scary Facts of Dairy Violate the Five Freedoms," *Psychology Today*, December 9, 2018, https://www.psychologytoday.com/us/blog/animal-emotions/201812/the-scary-facts-dairy-violate-the-five-freedoms; and Ashley Capps, "10 Dairy Facts the Industry Doesn't Want You to Know," Free from Harm, accessed September 29, 2023, https://freefromharm.org/dairyfacts. For information on going dairy-free, visit Switch4good.org.

p. 159 — *While it is very difficult to track accurately how many*: Hemi Kim, "Is Veganism Becoming More Popular? Using Data to Track the Growing Trend," *Sentient Media*, May 13, 2022, https://sentientmedia.org/increase-in-veganism.

p. 160 — *In his outstanding book* The Meat Paradox: Marc Bekoff, "Why Is Eating Meat So Emotionally and Ethically Challenging?," *Psychology Today*, February 19, 2022, https://www.psychologytoday.com/us/blog/animal-emotions/202202/why-is-eating-meat-so-emotionally-and-ethically-challenging.

p. 160 — *I try to capture the same thing when I tell people*: Marc Bekoff, "Babe, Lettuce, and Tomato: Dead Pig Walking," *Psychology Today*, October 9, 2011, https://www.psychologytoday.com/us/blog/animal-emotions/201110/babe-lettuce-and-tomato-dead-pig-walking.

p. 160 — *Consider the Kentucky Derby*: Details about the Kentucky Derby and the Santa Anita Racetrack are from the following: Bob D'Angelo, "Two More Horses Euthanized at Churchill Downs; Death Total at 7 before Kentucky Derby," *WHIOTV7*, May 6, 2023, https://www.whio.com/news/trending/sixth-horse-euthanized-churchill-downs-race-prior-149th-kentucky-derby/6ALN7ZOQKVECTBNO2VLEKW72BA; Elizabeth Banicki, "What Do Horses Feel at the Kentucky Derby? Mostly Fear and Pain," *The Guardian*, May 3, 2023, https://www.theguardian.com/sport/2023/may/03/kentucky-derby-racehorses-deaths-animal-welfare; Kathy Guillermo, "Four Dead Horses in Six Days at Churchill Downs Shows Horse Racing's Reality Problem," *Lexington Herald-Leader*, May 3, 2023, https://news.yahoo.com/four-dead-horses-six-days-142716137.html; Associated Press, "2 Horses Die from Injuries at Churchill Downs, Bringing Total to 12 at Home of Kentucky Derby," May 27, 2023, https://www.newsday.com/amp/sports/horse-racing/2-horses-die-from-injuries-at-churchill-downs-bringing-total-to-12-at-home-of-kentucky-derby-lev7il39; Merritt Clifton, "Necropsies Dominate the 2023 Kentucky Derby Post-Mortems," Animals

24-7, May 8, 2023, https://www.animals24-7.org/2023/05/08/necropsies-dominate-the-2023-kentucky-derby-post-mortems; and Jessica Ogilvie and Nate Perez, "Santa Anita Park Reports 6 Horse Deaths So Far in 2023, Renewing Racing Scrutiny," *LAist*, February 13, 2023, https://laist.com/news/santa-anita-park-reports-6-horse-deaths-so-far-in-2023-renewing-racing-scrutiny.

p. 161 *Calves are routinely and brutally abused*: Sylvia Dixon et al., "Behaviours Expressed by Rodeo Calves during Different Phases of Roping," *Animals* 13, no. 3 (January 2023), https://www.mdpi.com/2076-2615/13/3/343.

p. 161 *There also are rodeo-style chicken roping contests*: "United Poultry Concerns Condemns Chicken Roping Contest, Urges Compassionate Conduct," United Poultry Concerns, February 8, 2023, https://upc-online.org/nr/230208_upc_condemns_chicken_roping_contest_urges_compassionate_conduct.html. The quote is from Wendy Corr, "Yes, Chicken Roping in Wyoming Is a Real Thing, Moorcroft Holding 9th Annual Event in February," *Cowboy State Daily*, January 21, 2023, https://cowboystatedaily.com/2023/01/21/yes-chicken-roping-in-wyoming-is-a-real-thing-moorcroft-holding-9th-annual-event-in-february.

p. 161 *"But whatever thrill is to be derived from staring"*: Charles Siebert, "The Dark Side of Zootopia," *New York Times Magazine*, November 18, 2014, https://www.nytimes.com/2014/11/23/magazine/the-dark-side-of-zootopia.html.

p. 162 *"little to no systematic research has been conducted"*: AZA, "Visitor Learning in Zoos and Aquariums: A Literature Review," accessed September 29, 2023, https://www.informalscience.org/sites/default/files/AZA-Visitor_Learning_in_Zoos_Aquraiums_Literature_Review_0.pdf.

p. 162 *"It has never been more evident that conservation"*: "Staff" page, Association of Zoos & Aquariums, accessed September 29, 2023, https://www.aza.org/staff?locale=en.

p. 162 *"Any zoo that sits around and tells you"*: V. Croke, *The Modern Ark: The Story of Zoos: Past, Present, and Future* (New York: Scribner, 1997), 171.

p. 162 *many people worry that focusing on the role that zoos*: Liz Tyson-Griffin, "Animal Captivity Is a Dangerous Distraction from Real Conservation Efforts," Earth.org, August 17, 2023, https://earth.org/wild-animal-captivity.

p. 162 *Research published in the journal* Conservation Biology: Eric Jensen, "Evaluating Children's Conservation Biology Learning at the Zoo," *Conservation Biology* 28, no. 4 (August 2014), https://conbio.onlinelibrary.wiley.com/doi/abs/10.1111/cobi.12263; "Zoos Neither Educate Nor Empower Children," *Freedom for Animals*, April 10, 2019, https://www.freedomforanimals.org.uk/news/zoos-neither-educate-nor-empower-children; Marc Bekoff, "What Do Zoos Teach about Biodiversity and Does It Matter?," *Psychology Today*, March 8, 2014, https://www.psychologytoday.com/us/blog/animal-emotions/201403/what-do-zoos-teach-about-biodiversity-and-does-it-matter; and "Animals in Captivity: Do Zoos Actually Educate Visitors?," Penn State University, November 28, 2015, https://sites.psu.edu/siowfa15/2015/11/28/animals-in-captivity-do-zoos-actually-educate-visitors. In addition, see Andrew Moss, Eric Allen Jensen, and Markus Gusset, "A Global Evaluation of Biodiversity Literacy in Zoo and Aquarium Visitors," World Association of Zoos and Aquariums, January 2014, https://www.researchgate.net/publication/266444881_A_Global_Evaluation_of_Biodiversity_Literacy_in_Zoo_and_Aquarium_Visitors. The authors conclude, "Some positive evidence in this case would indicate knowing that biodiversity is related to biological phenomena with *no* evidence of understanding the breadth or variety of plant and animal species, the interdependency of species, the genetic value of biodiversity, the importance of biodiversity for humans or the need for biodiversity conservation" (my emphasis on *no*).

p. 162 *Although more people have shown an interest in conservation*: Gabriel Henrique de Oliveira

Endnotes

Caetano et al., "Evaluating Global Interest in Biodiversity and Conservation," *Conservation Biology* 37, no. 5 (October 2023), https://conbio.onlinelibrary.wiley.com/doi/10.1111/cobi.14100.

p. 163   *Many zoos engage in "conservation washing."*: Jessica Scott-Reid, "Beware the Rise of 'Conservation Washing' — The Latest Gambit to Keep Animals in Captivity," *Globe and Mail*, December 8, 2023, https://www.theglobeandmail.com/opinion/article-beware-the-rise-of-conservation-washing-the-latest-gambit-to-keep.

p. 163   *Captivity results in physical and behavioral changes*: Donna Lu, "Animals Bred in Captivity Develop Physical Changes That May Hinder Survival in the Wild, Research Finds," *The Guardian*, November 20, 2022, https://www.theguardian.com/australia-news/2022/nov/21/animals-bred-in-captivity-develop-physical-changes-that-may-hinder-survival-in-the-wild-research-finds; and Donna Lu, "Breeding Birds in Captivity May Alter Their Wing Shapes and Reduce Post-Release Survival Chances," *The Guardian*, March 18, 2023, https://www.theguardian.com/australia-news/2023/mar/19/breeding-birds-in-captivity-may-alter-their-wing-shapes-and-reduce-post-release-survival-chances.

p. 163   *Research shows large mammals suffer brain damage*: Marc Bekoff, "Research Shows Big Mammals Suffer Brain Damage in Captivity," *Psychology Today*, November 15, 2020, https://www.psychologytoday.com/us/blog/animal-emotions/202011/research-shows-big-mammals-suffer-brain-damage-in-captivity.

p. 163   *The results of an opinion poll taken in the UK*: "The People's Vote: No More Large Animals in Zoos," *Skem News*, February 16, 2023, https://skemnews.com/the-peoples-vote-no-more-large-animals-in-zoos.

p. 163   *And in February 2023, Puerto Rico announced*: Associated Press, "Puerto Rico Will Close Its Only Zoo after Long-Standing Complaints," *NBC News*, February 28, 2023, https://www.nbcnews.com/news/latino/puerto-rico-will-close-only-zoo-longstanding-complaints-rcna72787.

p. 163   *In June 2006, an elephant named Gita died*: Jeanne McDowell, "Are Zoos Killing Elephants?," *Time*, June 12, 2006.

p. 164   *I and my students studied coyotes living in Grand Teton National Park*: For a summary of my research on coyotes, see Marc Bekoff and M. C. Wells, "Social Behavior and Ecology of Coyotes," *Advances in the Study of Behavior* 16 (1986): 251–338.

p. 165   *Then there is a zoo in the Netherlands that once thoughtfully*: "Dutch Plan Orangutan Web Dating," *BBC News*, August 15, 2006, http://news.bbc.co.uk/2/hi/europe/4794279.stm.

p. 166   *In 2006, Suma, a forty-five-year-old elephant in the Zagreb Zoo*: "Mozart Soothes Bereaved Elephant," *ABC News*, June 29, 2006, https://www.abc.net.au/news/2006-06-30/mozart-soothes-bereaved-elephant/1790102.

p. 166   *Since 1991, more than thirty US zoos have closed*: "Closed or Closing Elephant Zoo Exhibits," In Defense of Animals, accessed October 5, 2023, https://www.idausa.org/campaign/elephants/closed-closing-elephant-zoo-exhibits.

p. 166   *In 2006, the* New York Times *reported that the Bronx Zoo*: Details about the Bronx Zoo and Patty and Happy come from the following: Charles Siebert, "An Elephant Crackup?," *New York Times Magazine*, October 8, 2006, http://www.nytimes.com/2006/10/08/magazine/08elephant.html; G. A. Bradshaw et al., "Elephant Breakdown," *Nature* 433 (2005): 807; Robbie Sequeira, "City Council Bill to Ban Elephant Captivity in NYC Puts Focus on Bronx Zoo's Happy and Patty," *Bronx Times*, March 16, 2023, https://www.bxtimes.com/city-council-ban-elephant-captivity-bronx-zoos; Jonathan Custodio, "Set Bronx Zoo Elephants Happy and Patty Free, Council Bill Orders," *The City*, March 16, 2023; and "Legal Fight to Free Happy the Elephant Continues in New York," Nonhuman Rights Project, accessed October 5, 2023, https://www.nonhumanrights.org/blog/fight-to-free-happy-continues.

p. 167 *In a 2018 article called "A Postzoo Future"*: Jessica Pierce and Marc Bekoff, "A Postzoo Future: Why Welfare Fails Animals in Zoos," *Journal of Applied Animal Welfare Science* 21, no. 1 (October 2018), https://www.tandfonline.com/doi/full/10.1080/10888705.2018.1513838.

p. 167 *These prescriptions would also apply to captive animals in aquariums*: In November 2022, the Jane Goodall Institute put forth their "Statement on Cetacean Captivity" (https://www.thejanegoodallinstitute.com/jgi-statement-on-cetacean-captivity), which summarizes their stance as follows: "The Jane Goodall Institute calls for an immediate worldwide, permanent ban on capturing, keeping and breeding cetaceans in captivity." For more information on this issue, visit the Whale Sanctuary Project (https://whalesanctuaryproject.org) and watch the wonderful 2013 documentary *Blackfish*, which focuses on Tilikum, an orca who killed three people. See also Marc Bekoff, "Do Orcas Go Crazy Because of Petting Pools and False Hopes?," *Psychology Today*, March 30, 2015, https://www.psychologytoday.com/us/blog/animal-emotions/201503/do-orcas-go-crazy-because-petting-pools-and-false-hopes.

p. 167 *As Dale Jamieson aptly puts it, "Zoos deny agency"*: Dale Jamieson, "Sentience Is More Complicated Than You Think," Law, Ethics, & Animals Program at Yale Law School, February 17, 2022, https://www.youtube.com/watch?v=iGIeyqY1Rgo.

p. 167 *For all beings, agency and consent have positive effects*: Marc Bekoff, "Why Consenting Cats Are Happier Cats," *Psychology Today*, March 12, 2023, https://www.psychologytoday.com/us/blog/animal-emotions/202303/consenting-cats-are-happier-cats.

p. 167 *Bernard Harrison, past director of the Singapore Zoo*: Amanda Chai, "Catching Up with Bernard Harrison, Singapore's Very Own Zoo Tycoon," *SG Magazine*, August 23, 2017, https://sgmagazine.com/people/news/bernard-harrison-singapore-zoo-tycoon.

p. 168 *"Unfortunately, the bulk of zoos in existence today"*: Jenny Gray, *Zoo Ethics: The Challenges of Compassionate Conservation* (Ithaca, NY: Comstock Publishing Associates, 2017), 208. Also, In Defense of Animals published their "10 Worst Zoos for Elephants 2022" (https://www.idausa.org/campaign/elephants/10-worst-zoos-for-elephants-2022), which includes major institutions like the Oregon Zoo, Kansas City Zoo, Houston Zoo, Zoo Atlanta, and Cincinnati Zoo.

p. 168 *Many people don't know that zoos routinely kill*: Ben Tobias and Paul Kirby, "Sweden Furuvik Zoo: Anger over Shooting of Chimpanzees in Zoo Escape," *BBC News*, December 16, 2022, https://www.bbc.com/news/world-europe-64005819.

p. 168 *"does not publish these records or advertise the number"*: Hannah Barnes, "How Many Healthy Animals Do Zoos Put Down?," *BBC News*, February 27, 2014, https://www.bbc.com/news/magazine-26356099.

p. 168 *Nevertheless, cases of zoothanasia garner attention*: The details in this paragraph come from the following: Laura Smith-Spark and Bharati Naik, "Copenhagen Zoo Kills 4 Lions, Weeks after Shooting Giraffe," *CNN*, March 26, 2104, https://www.cnn.com/2014/03/26/world/europe/copenhagen-zoo-lions/index.html; Marc Bekoff, "Swedish Zoo 'Zoothanizes' Nine Healthy, 'Useless' Lion Cubs," *Psychology Today*, January 16, 2018, https://www.psychologytoday.com/us/blog/animal-emotions/201801/swedish-zoo-zoothanizes-nine-healthy-useless-lion-cubs; Jon Henley, "Three Chimpanzees Shot Dead after Escape from Swedish Zoo," *The Guardian*, December 14, 2022, https://www.theguardian.com/world/2022/dec/14/three-chimpanzees-shot-dead-after-escape-from-swedish-zoo; Mehek Mazhar, "'Surreal' Killing of 4 Chimpanzees Has Zoologist Reconsidering Ties with Swedish Zoo," *CBC*, December 21, 2022, https://www.cbc.ca/radio/asithappens/sweden-zoo-chimpanzees-killed-1.6691262; Erika Helms, "Jane Goodall Institute Statement on the Killing of Chimpanzees in Swedish EAZA Zoo Furuviksparken," Jane Goodall Institute, December 19, 2022, https://www

.thejanegoodallinstitute.com/statement-on-the-killing-of-chimpanzees-in-swedish-eaza-zoo-furuviksparken; "Dundee Zoo 'Heartbroken' after Wolf Pack Euthanised," *BBC News*, March 28, 2023, https://www.bbc.com/news/uk-scotland-tayside-central-65052453; and Marc Bekoff, "Dundee Zoo Needlessly Kills a Wolf Pack After Their Leader Dies," *Marc Bekoff, Ph.D.* (blog), April 6, 2023, https://marcbekoff.com/marcs-essays/f/dundee-zoo-needlessly-kills-a-wolf-pack-after-their-leader-dies.

p. 169   *For captive pandas, this amounts to a breeding boondoggle*: Marc Bekoff, "Pandering Pandas: Sweetie and Sunshine on Loan in Edinburgh," *Psychology Today*, December 12, 2011, https://www.psychologytoday.com/us/blog/animal-emotions/201112/pandering-pandas-sweetie-and-sunshine-loan-in-edinburgh; Marc Bekoff, "The Captive Panda Breeding Boondoggle: The Invisible Side," *Psychology Today*, September 5, 2019, https://www.psychologytoday.com/us/blog/animal-emotions/201909/the-captive-panda-breeding-boondoggle-the-invisible-side; and Marc Bekoff, "Cuddly Pandas Are 'Cuteness Crack' Says British Zoologist," *Psychology Today*, April 9, 2018, https://www.psychologytoday.com/us/blog/animal-emotions/201804/cuddly-pandas-are-cuteness-crack-says-british-zoologist.

p. 169   *This happened in early 2023 at the Basel Zoo*: J. A. Allen, "Zoo Basel Puts Baby Orangutan to Sleep — Now a Shitstorm Is Raging," *Then24*, February 1, 2023, https://then24.com/2023/02/01/zoo-basel-puts-baby-orangutan-to-sleep-now-a-shitstorm-is-raging; and Erika Helms, "JGI Statement on the Death of a Baby Orangutan in Switzerland," Jane Goodall Institute, February 6, 2023, https://www.thejanegoodallinstitute.com/jgi-statement-on-the-death-of-a-baby-orangutan-in-switzerland.

p. 170   *"We are part of a generation of shame"*: This quote by Jill Robinson is from a personal communication with the author.

p. 171   *Y2Y is establishing a system of corridors*: For more information on the Yellowstone to Yukon Conservation Initiative, visit www.y2y.net.

p. 171   *Numerous wild animals are killed by federal agencies*: The statistics in this paragraph come from the following: Oliver Milman, "'A Barbaric Federal Program': US Killed 1.75m Animals Last Year — or 200 Per Hour," *The Guardian*, March 25, 2022, https://www.theguardian.com/world/2022/mar/25/us-government-wildlife-services-animals-deaths; Dina Fine Maron, "US Government Agency Accidentally Killed Almost 3,000 Animals in 2021," *National Geographic*, April 12, 2022, https://www.nationalgeographic.com/animals/article/us-government-agency-accidentally-killed-almost-3000-animals-in-2021; and Colin Tiernan, "'The Hired Gun of the Livestock Industry': Wildlife Services, a Federal Agency, Kills More Than 100,000 Animals in Washington Every Year," *Spokesman-Review*, March 12, 2023, https://www.spokesman.com/stories/2023/mar/12/the-hired-gun-of-the-livestock-industry-this-feder. For more general information on wildlife issues, visit the Humane Society of the United States, https://www.humanesociety.org.

p. 172   *"Officials acknowledge that killing deer and elk"*: "Colorado Quits Killing Elk, Deer to Contain Disease," *Casper Star Tribune*, March 31, 2006, http://www.casperstartribune.net/articles/2006/04/02/news/regional/5e3061a2180d38868725714100749d91.txt.

p. 173   *As Bethany Brookshire shows in her 2022 book*: Marc Bekoff, "Calling Animals 'Pests' Is More About Us Than Them," *Psychology Today*, December 6, 2022, https://www.psychologytoday.com/us/blog/animal-emotions/202212/calling-animals-pests-is-more-about-us-them.

p. 173   *Wildlife and conservation scientist Dr. Gosia Bryja aptly notes*: Gosia Bryja, "Language and the Knife: Silencing Nature," *Medium*, March 31, 2023, https://gbryja.medium.com/language-and-the-knife-silencing-nature-5ac36a2378cf.

p. 174   *I know because people have told me that this small change*: Marc Bekoff, "As Food Animals

Became 'Things,' Their Feelings Were Ignored," *Psychology Today*, February 22, 2022, https://www.psychologytoday.com/us/blog/animal-emotions/202202/food-animals-became-things-their-feelings-were-ignored.

p. 175 *What is it other than murder when zoos "euthanize"*: Marc Bekoff, "Murdering Animals: A Book about Social and Species Justice," *Psychology Today*, April 10, 2018, https://www.psychologytoday.com/us/blog/animal-emotions/201804/murdering-animals-book-about-social-and-species-justice.

p. 175 *For instance, take the phrase "ventilation shutdown"*: "Ventilation Shutdown Used to 'Depopulate' Farm Animals During Pandemic Causes Severe Suffering," Animal Welfare Institute, July 1, 2020, https://awionline.org/press-releases/ventilation-shutdown-used-depopulate-farm-animals-during-pandemic-causes-severe. See also Veterinarians Against Ventilation Shutdown, https://www.vavsd.org/about.

p. 175 *Dr. Karen Davis, president and founder of United Poultry Concerns*: Karen Davis, "Do Chickens Mind Seeing Other Chickens Killed in Their Presence?," United Poultry Concerns, March 10, 2023, https://upc-online.org/podcasts/230310_do_chickens_mind_seeing_other_chickens_killed.html.

p. 176 *"the ethical, sportsmanlike, and lawful pursuit"*: "Fair Chase Statement," Boone and Crockett Club, accessed September 30, 2023, https://www.boone-crockett.org/fair-chase-statement.

p. 176 *"Fair Chase rules make sure hunters have no unfair"*: "Fair Chase," Hunter-ed.com, accessed September 30, 2023, https://www.hunter-ed.com/alaska/studyGuide/Fair-Chase/201001_86886.

p. 177 *Compassionate conservation is based on the ethical position*: For more information on compassionate conservation, see the following: "What Is Compassionate Conservation?," University of Technology, Sydney (UTS), accessed September 30, 2023, https://www.uts.edu.au/research-and-teaching/our-research/centre-compassionate-conservation/about-us/what-compassionate-conservation; Daniel Ramp and Marc Bekoff, "Compassion as a Practical and Evolved Ethic for Conservation," *BioScience* 65, no. 3, (March 2015), https://academic.oup.com/bioscience/article/65/3/323/236866; and Arian Wallach et al., "Summoning Compassion to Address the Challenges of Conservation," *Conservation Biology* 32, no. 6 (April 2018), https://www.researchgate.net/publication/324801356_Summoning_compassion_to_address_the_challenges_of_conservation.

p. 177 *Kristy Ferraro and her colleagues correctly argue*: Kristy Ferraro et al., "Revisiting Two Dogmas of Conservation Science," *Conservation Biology* 37, no. 4 (April 2023), https://conbio.onlinelibrary.wiley.com/doi/10.1111/cobi.14101?af=R.

p. 177 *Critics of compassionate conservation argue*: Marc Bekoff, "Compassionate Conservation Isn't Seriously or Fatally Flawed," *Psychology Today*, May 20, 2019, https://www.psychologytoday.com/us/blog/animal-emotions/201905/compassionate-conservation-isnt-seriously-or-fatally-flawed.

p. 177 *the United Nations Harmony with Nature Knowledge Network*: To learn more about this program, visit http://www.harmonywithnatureun.org/welcome.

p. 178 *"Wildlife SOS spearheaded a conservation success story"*: Marc Bekoff, "Compassionate Conservation Meets Cecil the Slain Lion," *Psychology Today*, August 9, 2015, https://www.psychologytoday.com/us/blog/animal-emotions/201508/compassionate-conservation-meets-cecil-the-slain-lion.

p. 179 *"Changing the focus from 'problem animals'"*: "Leopard Landscapes: Coexisting with Carnivores in Countryside and City," *Economic & Political Weekly* 50, no. 1 (January 3, 2015), https://www.epw.in/journal/2015/1/reports-states-web-exclusives/leopard-landscapes.html-0.

Endnotes

p. 179 *What's pertinent here is that emotions can be a source*: Marc Bekoff, "Do Emotions 'Get in the Way' of Progress in Conservation?," *Psychology Today*, April 16, 2021, https://www.psychologytoday.com/us/blog/animal-emotions/202104/do-emotions-get-in-the-way-progress-in-conservation; and Chelsea Batavia et al., "Emotion as a Source of Moral Understanding in Conservation," *Conservation Biology* 35, no. 5 (October 2021): 1380–87, https://conbio.onlinelibrary.wiley.com/doi/abs/10.1111/cobi.13689.

p. 179 *This was brought home to me when I read a January 2023 essay*: Kira Cassidy et al., "Human-Caused Mortality Triggers Pack Instability in Gray Wolves," *Frontiers in Ecology and the Environment* (January 2023), https://esajournals.onlinelibrary.wiley.com/doi/10.1002/fee.2597; and Marc Bekoff, "Wolf Packs Suffer When Humans Kill Their Leaders," *Psychology Today*, 2023, https://www.psychologytoday.com/us/blog/animal-emotions/202301/wolf-packs-suffer-when-humans-kill-their-leaders.

p. 180 *Philosopher Martha Nussbaum has argued that justice*: Marc Bekoff, "Justice for Animals Means Freedom to Do What Comes Naturally," *Psychology Today*, December 12, 2022, https://www.psychologytoday.com/us/blog/animal-emotions/202212/justice-animals-means-freedom-do-what-comes-naturally.

p. 181 *As humane educator Zoe Weil aptly notes*: Zoe Weil, "The World Becomes What You Teach," TEDx Talk, January 14, 2011, https://www.youtube.com/watch?v=t5HEV96dIuY.

Afterword: Compassion and Justice for All

p. 183 *"Life supports life. Animals are the key"*: Keggie Carew, "Animals Can Save Us — If We Let Them," *Time*, April 22, 2023, https://time.com/6272815/animals-can-save-us-earth-day.

p. 183 *It can be true that other animals often suffer in silence*: David Grimm, "Suffering in Silence," *Science*, March 10, 2023, https://www.sciencemagazinedigital.org/sciencemagazine/library/item/10_march_2023/4085716.

p. 183 *"I think sometimes we need to take a step back"*: Kaleigh Dray, "Sir David Attenborough's Most Profound Quotes on Wanderlust, Women's Rights, and Mother Nature," *Stylist*, https://www.stylist.co.uk/entertainment/celebrity/best-david-attenborough-quotes/33686; and Edwin Page, *The Vegan Way of Life* (independently published, 2021).

p. 184 *"The One Health approach is a way of looking at the world"*: Marc Bekoff, "Why People Should Care About Animal and Human Suffering," *Psychology Today*, December 1, 2018, https://www.psychologytoday.com/us/blog/animal-emotions/201812/why-people-should-care-about-animal-and-human-suffering.

p. 184 *"Caring for nonhumans, for their own sake"*: John A. Vucetich et al., "Evaluating Whether Nature's Intrinsic Value Is an Axiom of or Anathema to Conservation," *Conservation Biology* 29, no. 2 (April 2015), https://conbio.onlinelibrary.wiley.com/doi/abs/10.1111/cobi.12464.

p. 185 *Theologian Stephen Scharper resolves this contradiction*: Stephen Scharper, *Redeeming the Time* (New York: Theommes Continuum, 1997), 188.

p. 189 *We must make kind and humane choices*: This is a point stressed by many people. For numerous easy ways to do this, see Ingrid Newkirk, *Making Kind Choices: Everyday Ways to Enhance Your Life through Earth- and Animal-Friendly Living* (New York: St. Martin's Griffin, 2005). See also *PETA Global* magazine (https://petaglobal.cld.bz), which features animal rights news, vegan recipes, and practical ways to help animals.

Bibliography

Alcock, John. *Animal Behavior: An Evolutionary Approach.* 8th ed. Sunderland, MA: Sinauer Associates, Inc., 2005.

Allen, Colin, and Marc Bekoff. *Species of Mind: The Philosophy and Biology of Cognitive Ethology.* Cambridge, MA: MIT Press, 1997.

Allen, Moira. *Laugh-Out-Loud Victorian Poetry: Animal Antics.* VictorianVoices.Net, 2019.

Appleby, M. C., J. A. Mench, and B. O. Hughes. *Poultry Behaviour and Welfare.* Cambridge, MA: CABI Publishing, 2004.

Archer, John. *The Nature of Grief: The Evolution and Psychology of Reactions to Loss.* New York: Routledge, 1999.

Balcombe, Jonathan. *Pleasurable Kingdom: Animals and the Nature of Feeling Good.* London: Macmillan, 2006.

———. *The Use of Animals in Education: Problems, Alternatives, and Recommendations.* Washington, DC: Humane Society of the United States, 2000.

Bateson, P. P. G. "Assessment of Pain in Animals." *Animal Behaviour* 42 (1991): 827–39.

Becker, Marty. *The Healing Power of Pets.* New York: Hyperion, 2002.

Bekoff, Marc. "Animal Emotions and Animal Sentience and Why They Matter: Blending 'Science Sense' with Common Sense, Compassion and Heart." In *Animals, Ethics, and Trade.* Edited by J. Turner and J. D'Silva, 27–40. London: Earthscan Publishing, 2006.

———. "Animal Emotions: Exploring Passionate Natures." *BioScience* 50 (2000): 861–70.

———. *The Animal Manifesto: Six Reasons for Expanding Our Compassion Footprint.* Novato, CA: New World Library, 2010.

———. "Animal Passions and Beastly Virtues: Cognitive Ethology as the Unifying Science for Understanding the Subjective, Emotional, Empathic, and Moral Lives of Animals." *Zygon: Journal of Religion and Science* 41 (2006): 71–104.

Bibliography

———. *Animal Passions and Beastly Virtues: Reflections on Redecorating Nature.* Philadelphia: Temple University Press, 2006.
———. *Canine Confidential: Why Dogs Do What They Do.* Chicago: University of Chicago Press, 2018.
———. "The Communication of Play Intention: Are Play Signals Functional?" *Semiotica* 15 (1975): 231–39.
———. *Dogs Demystified: An A-Z Guide to All Things Canine.* Novato, CA: New World Library, 2023.
———. ed. *Encyclopedia of Animal Behavior.* Westport, CT: Greenwood Publishing Group, 2004.
———. ed. *Encyclopedia of Animal Rights and Animal Welfare.* Westport, CT: Greenwood Publishing Group, 1998.
———. ed. *Encyclopedia of Human-Animal Relationships: A Global Exploration of Our Connections with Animals.* Westport, CT: Greenwood Publishing Group, 2007.
———. ed. *Ignoring Nature No More: The Case for Compassionate Conservation.* Chicago: University of Chicago Press, 2013.
———. *Minding Animals: Awareness, Emotions, and Heart.* New York: Oxford University Press, 2002.
———. "Play Signals as Punctuation: The Structure of Social Play in Canids." *Behaviour* 132 (1995): 419–29.
———. "The Public Lives of Animals: A Troubled Scientist, Pissy Baboons, Angry Elephants, and Happy Hounds." *Journal of Consciousness Studies* 13 (2006): 115–31.
———. *Rewilding Our Hearts: Building Pathways of Compassion and Coexistence.* Novato, CA: New World Library, 2010.
———. ed. *The Smile of a Dolphin: Remarkable Accounts of Animal Emotions.* New York: Random House / Discovery Books, 2000.
———. "Social Communication in Canids: Evidence for the Evolution of a Stereotyped Mammalian Display." *Science* 197 (1977): 1097–99.
———. *Strolling with Our Kin: Speaking for and Respecting Voiceless Animals.* New York: Lantern Books, 2000.
———. "Wild Justice and Fair Play: Cooperation, Forgiveness, and Morality in Animals." *Biology & Philosophy* 19 (2004): 489–520.
Bekoff, Marc, and Colin Allen. "Cognitive Ethology: Slayers, Skeptics, and Proponents." In *Anthropomorphism, Anecdote, and Animals.* Edited by R. W. Mitchell, N. Thompson, and L. Miles, 313–34. Albany, NY: SUNY Press, 1997.
Bekoff, Marc, Colin Allen, and Gordon M. Burghardt, eds. *The Cognitive Animal: Empirical and Theoretical Perspectives on Animal Cognition.* Cambridge, MA: MIT Press, 2002.
Bekoff, Marc, and John A. Byers, eds. *Animal Play: Evolutionary, Comparative, and Ecological Perspectives.* New York: Cambridge University Press, 1998.
Bekoff, Marc, and Dale Jamieson. "Ethics and the Study of Carnivores: Doing Science While Respecting Animals." In *Carnivore Behavior, Ecology, and Evolution.* Edited by J. Gittleman, 16–45. Ithaca, NY: Cornell University Press, 1996.

———. eds. *Readings in Animal Cognition*. Cambridge, MA: MIT Press, 1996.

———. "Reflective Ethology, Applied Philosophy, and the Moral Status of Animals." *Perspectives in Ethology* 9 (1991): 1–47.

Bekoff, Marc, and Jessica Pierce. *The Animals' Agenda: Freedom, Compassion, and Coexistence in the Human Age*. Boston: Beacon Press, 2017.

———. *Unleashing Your Dog: A Field Guide to Giving Your Canine Companion the Best Life Possible*. Novato, CA: New World Library, 2019.

———. *Wild Justice: The Moral Lives of Animals*. Chicago: University of Chicago Press, 2009.

Berns, Gregory. *What It's Like to Be a Dog: And Other Adventures in Animal Neuroscience*. New York: Basic Books, 2017.

Berry, Thomas. *The Great Work: Our Way into the Future*. New York: Bell Tower, 1999.

Brookshire, Bethany. *Pests: How Humans Create Animal Villains*. New York: HarperCollins, 2022.

Buchmann, Stephen. *What a Bee Knows: Exploring the Thoughts, Memories, and Personalities of Bees*. Washington, DC: Island Press, 2023.

Burghardt, Gordon M. "Amending Tinbergen: A Fifth Aim for Ethology." In *Anthropomorphism, Anecdote, and Animals*. Edited by R. W. Mitchell, N. Thompson, and L. Miles, 254–76. Albany, NY: SUNY Press, 1997.

———. *The Genesis of Play*. Cambridge, MA: MIT Press, 2005.

Campbell, Jeff. *Daisy to the Rescue: True Stories of Daring Dogs, Paramedic Parrots, and Other Animal Heroes*. San Francisco: Zest Books, 2014.

Calvo, Paco, and Natalie Lawrence. *Planta Sapiens: Unmasking Plant Intelligence*. New York: W. W. Norton & Company, 2023.

Carbone, Larry. *What Animals Want: Expertise and Advocacy in Laboratory Animal Welfare Policy*. New York: Oxford University Press, 2004.

Challenger, Melanie, ed. *Animal Dignity: Philosophical Reflections on Non-Human Existence*. New York: Bloomsbury Academic, 2023.

Cheney, Dorothy L., and Robert M. Seyfarth. *How Monkeys See the World: Inside the Mind of Another Species*. Chicago: University of Chicago Press, 1990.

Cooke, Lucy. *Bitch: On the Female of the Species*. New York: Basic Books, 2022.

Coulter, Kendra. *Defending Animals: Finding Hope on the Front Lines of Animal Protection*. Cambridge, MA: MIT Press, 2023.

Damasio, Antonio. *Descartes' Error: Emotion, Reason, and the Human Brain*. New York: Penguin, 2005.

———. *The Feeling of What Happens: Body and Emotion in the Making of Consciousness*. New York: Harcourt Brace, 1999.

Danovich, Tove. *Under the Henfluence: Inside the World of Backyard Chickens and the People Who Love Them*. Evanston, IL: Agate Publishing, 2023.

Darwin, Charles. *The Descent of Man and Selection in Relation to Sex*. New York: Random House, 1871/1936.

———. *The Expression of the Emotions in Man and Animals*, 3rd ed. New York: Oxford University Press, 1872/1998.

Bibliography

Davis, Karen. *The Holocaust and the Henmaid's Tale: A Case for Comparing Atrocities.* New York: Lantern Books, 2005.
Dawkins, Marian S. *Through Our Eyes Only?* New York: Oxford University Press, 1992.
de Waal, Frans. *Are We Smart Enough to Know How Smart Animals Are?* New York: W. W. Norton & Company, 2016.
———. *Good-Natured: The Origins of Right and Wrong in Humans and Other Animals.* Cambridge, MA: Harvard University Press, 1996.
———. *Mama's Last Hug: Animal Emotions and What They Tell Us about Ourselves.* New York: W. W. Norton & Company, 2019.
———. *Our Inner Ape.* New York: Riverhead, 2005.
———. *Primates and Philosophers: How Morality Evolved.* Princeton, NJ: Princeton University Press, 2006.
DeYoung, Sarah E., and Ashley K. Farmer. *All Creatures Safe and Sound: The Social Landscape of Pets in Disasters.* Philadelphia: Temple University Press, 2021.
Dodman, Nicholas. *The Cat Who Cried for Help: Attitudes, Emotions, and the Psychology of Cats.* New York: Bantam, 1999.
———. *If Only They Could Speak: Stories about Pets and Their People.* New York: W. W. Norton & Company, 2002.
Doody, J. Sean, Vladimir Dinets, and Gordon M. Burghardt. *The Secret Social Lives of Reptiles.* Baltimore: Johns Hopkins University Press, 2021.
Drickamer, Lee C., Stephen H. Vessey, and Douglas Meikle. *Animal Behavior: Mechanisms, Ecology, and Evolution.* Dubuque, IA: William C. Brown Publishers, 2001.
D'Silva, Joyce. *Animal Welfare in World Religion.* New York: Routledge Taylor & Francis Group, 2023.
Dugatkin, Lee A. *Principles of Animal Behavior.* New York: W. W. Norton & Company, 2003.
Dugatkin, Lee A., and Marc Bekoff. "Play and the Evolution of Fairness: A Game Theory Model." *Behavioural Processes* 60 (2003): 209–14.
Duncan, I. J. H. "Poultry Welfare: Science or Subjectivity?" *British Poultry Science* 43 (2002): 643–52.
Dutcher, Jim, and Jamie Dutcher. *Wolves at Our Door.* New York: Pocket Books, 2002.
Eibl-Eibesfeldt, Irenäus. *Ethology.* New York: Holt, Rinehart & Winston, 1975.
Eisnitz, Gale A. *Slaughterhouse.* New York: Prometheus, 1997.
Essig, Mark. *Lesser Beasts: A Snout-to-Tail History of the Humble Pig.* New York: Basic Books, 2015.
Fagen, Robert. *Animal Play Behavior.* New York: Oxford University Press, 1981.
Fitzner, Zach. *Tears for Crocodilia: Evolution, Ecology, and the Disappearance of One of the World's Most Ancient Animals.* Yardley, PA: Westholme Publishing, 2022.
Flack, J. C., and Frans de Waal. "Any Animal Whatever: Darwinian Building Blocks of Morality in Monkeys and Apes." *Journal of Consciousness Studies* 7 (2000): 1–29.
Fox, Michael W. *Behaviour of Wolves, Dogs, and Related Canids.* London: Jonathan Cape, 1971.

———. *Bringing Life to Ethics: Global Bioethics for a Humane Society.* Albany, NY: SUNY Press, 2001.

———. *Eating with Conscience.* Troutdale, OR: New Sage Press, 1997.

Fradkin, Philip. *A River No More.* Berkeley: University of California Press, 1996.

Francione, Gary L. *Introduction to Animal Rights: Your Child or the Dog?* Philadelphia: Temple University Press, 2000.

Gigliotti, Carol. *The Creative Lives of Animals.* New York: New York University Press, 2022.

Goodall, Jane. *Through a Window.* Boston: Houghton-Mifflin, 1990.

Goodall, Jane, and Marc Bekoff. *The Ten Trusts: What We Must Do to Care for the Animals We Love.* San Francisco: HarperCollins, 2002.

Goodall, Jane, with Gary McAvoy and Gail Hudson. *Harvest for Hope: A Guide to Mindful Eating.* New York: Warner Books, 2005.

Greek, C. Ray, and Jean S. Greek. *Sacred Cows and Golden Geese: The Human Cost of Experiments on Animals.* New York: Theommes Continuum, 2000.

Gregg, Justin. *If Nietzsche Were a Narwhal: What Animal Intelligence Reveals about Human Stupidity.* Boston: Little Brown and Company, 2022.

Griffin, Donald R. *Animal Minds.* Chicago: University of Chicago Press, 1992.

———. *The Question of Animal Awareness: Evolutionary Continuity of Mental Experience.* New York: Rockefeller University Press, 1976/1981.

Gruen, Lori. *Ethics and Animals: An Introduction.* 2nd ed. New York: Cambridge University Press, 2021.

Gruen, Lori, and Justin Marceau, eds. *Carceral Logics: Human Incarceration and Animal Captivity.* New York: Cambridge University Press, 2022.

Hare, Brian, and Vanessa Woods. *Survival of the Friendliest: Understanding Our Origins and Rediscovering Our Common Humanity.* New York: Penguin Random House, 2021.

Hauser, Marc. *Wild Minds.* New York: Henry Holt, 1999.

Heinrich, Bernd. *Mind of the Raven: Investigations and Adventures with Wolf-Birds.* New York: Cliff Street Books, 1999.

Heister, Anja. *Beyond the North American Model of Wildlife Conservation: From Lethal to Compassionate Conservation.* London: Palgrave Macmillan, 2022.

Higgins, Jackie. *Sentient: How Animals Illuminate the Wonder of Our Human Senses.* New York: Simon & Schuster, 2022.

Hinde, Robert A. *Why Good Is Good: The Sources of Morality.* New York: Routledge, 2002.

Holland, Jennifer. *Unlikely Friendships: 47 Remarkable Stories from the Animal Kingdom.* New York: Workman Publishing Company, 2011.

———. *Unlikely Loves: 43 Heartwarming True Stories from the Animal Kingdom.* New York: Workman Publishing Company, 2013.

Hoyle, Richard, and Anita Krajnc, eds. *The Secret Life of Pigs: Stories of Compassion and the Animal Save Movement.* New York: Lantern, 2022.

Huffman, M. A. "Current Evidence for Self-Medication in Primates: A Multidisciplinary Perspective." *Yearbook of Physical Anthropology* 40 (1997): 171–200.

———. "Self-Medicative Behavior in the African Great Apes: An Evolutionary Perspective into the Origins of Human Traditional Medicine." *BioScience* 51 (2001): 651–61.

Hutton, Vicki. *Recognising and Responding to Animal Emotion in a Shared World*. Boca Raton, FL: CRC Press, 2023.

Irvine, Leslie. *If You Tame Me: Understanding Our Connections with Animals*. Philadelphia: Temple University Press, 2004.

Jamieson, Dale, and Marc Bekoff. "On Aims and Methods of Cognitive Ethology." *PSA: Proceedings of the Biennial Meeting of the Philosophy of Science Association* 1992 (1992), 110–24.

Jane Goodall Institute. *Local Voices, Local Choices: The Tacare Approach to Community-Led Conservation*. Redlands, CA: Esri Press, 2022.

Johnson, Lawrence. *A Morally Deep World: An Essay on Moral Significance and Environmental Ethics*. New York: Cambridge University Press, 1991.

Jordan, William. *A Cat Named Darwin: Embracing the Bond Between Man and Pet*. Boston: Mariner Books, 2003.

Kaplan, Gisela. *Bird Minds: Cognition and Behaviour of Australian Native Birds*. Clayton, Australia: CSIRO Publishing, 2015.

Kassam, Ammir, and Laila Kassam, eds. *Rethinking Food and Agriculture: New Ways Forward*. Sawaton, UK: Woodhead Publishing, 2020.

King, Barbara J. *How Animals Grieve*. Chicago: University of Chicago Press, 2013.

Kropotkin, Peter. *Mutual Aid: A Factor of Evolution*. Boston: Expanding Horizons Press, 1914.

Leese, Emilia A., and Eva J. Charalambides, eds. *Think Like a Vegan: What Everyone Can Learn from Vegan Ethics*. London: Unbound, 2022.

Lehner, Philip N. *Handbook of Ethological Methods*. New York: Cambridge University Press, 1996.

Levinson, Boris. *Pet-Oriented Child Psychotherapy*. Springfield, IL: Charles C. Thomas, 1969.

———. *Pets and Human Development*. Springfield, IL: Charles C. Thomas, 1972.

Leyhausen, Paul. *Cat Behavior: The Predatory and Social Behavior of Domestic and Wild Cats*. New York: Garland, 1978.

Livingston, John A. *The Fallacy of Wildlife Conservation*. Toronto: McClelland and Stewart, 1981.

Long, William J. *Brier-Patch Philosophy by "Peter Rabbit."* Boston and London: Ginn & Company, 1906.

Lorenz, Konrad Z. *Here I Am — Where Are You?* New York: Harcourt Brace Jovanovich, 1991.

MacLean, Paul. *The Triune Brain in Evolution: Role in Paleocerebral Functions*. New York: Plenum, 1970.

Marcus, Erik. *Meat Market: Animals, Ethics, and Money*. Newfield, NY: Brio Press, 2005.

Margodt, Koen. *The Welfare Ark: Suggestions for a Renewed Policy for Zoos*. Amsterdam: Vu University Press, 2001.

Marzluff, John, and Tony Angell. *Gifts of the Crow: How Perception, Emotion, and Thought Allow Smart Birds to Behave Like Humans*. New York: Atria, 2013.

Masson, Jeffrey Moussaieff. *The Nine Emotional Lives of Cats: A Journey into the Feline Heart*. New York: Ballantine Books, 2004.

Masson, Jeffrey Moussaieff, and Susan McCarthy. *When Elephants Weep: The Emotional Lives of Animals*. New York: Delacorte Press, 1995.

Matsuzawa, Tetsuro, ed. *Primate Origins of Human Cognition and Behavior*. New York: Springer, 2001.

Mbatha, Sicelo, and Bridget Pitt. *Black Lion: Alive in the Wilderness*. Johannesburg: Jonathan Bull Publishers, 2021.

McArthur, Jo-Anne, and Keith Wilson. *Hidden: Animals in the Anthropocene*. New York: Lantern Publishing and Media, 2020.

McCarthy, Susan. *Becoming a Tiger: How Baby Animals Learn to Live in the Wild*. New York: Harper Perennial, 2005.

McMillan, Franklin D., with Kathryn Lance. *Unlocking the Animal Mind: How Your Pet's Feelings Hold the Key to His Health and Happiness*. Emmaus, PA: Rodale, 2004.

Merskin, Debra L. *Seeing Species: Re-presentations of Animals in Media & Popular Culture*. Bern, Switzerland: Peter Lang, 2018.

Midkiff, Ken. *The Meat You Eat*. New York: St. Martin's Griffin, 2004.

Mock, Douglas W., and Geoffrey A. Parker. *The Evolution of Sibling Rivalry*. New York: Oxford University Press, 1997.

Moss, Cynthia. *Elephant Memories: Thirteen Years in the Life of an Elephant Family*. Chicago: University of Chicago Press, 2000.

Nader, Ralph. *Animal Envy: A Fable*. New York: Seven Stories Press, 2016.

Newkirk, Ingrid. *Making Kind Choices: Everyday Ways to Enhance Your Life Through Earth- and Animal-Friendly Living*. New York: St. Martin's Griffin, 2005.

Ohlson, Kristin. *Sweet in Tooth and Claw: Stories of Generosity and Cooperation in the Natural World*. Manchester, VT: Patagonia, 2022.

Panksepp, Jaak. "Affective Consciousness: Core Emotional Feelings in Animals and Humans." *Consciousness and Cognition* 14 (2005): 30–80.

———. *Affective Neuroscience*. New York: Oxford University Press, 1998.

Percival, Rob. *The Meat Paradox: Eating, Empathy, and the Future of Meat*. New York: Simon & Schuster, 2022.

Peterson, Dale. *Eating Apes*. Berkeley: University of California Press, 2003.

Pierce, Jessica, and Marc Bekoff. *A Dog's World: Imagining the Lives of Dogs in a World without Humans*. Princeton: Princeton University Press, 2021.

Pollan, Michael. *The Omnivore's Dilemma*. New York: Penguin Press, 2006.

Poole, Joyce. *Coming of Age with Elephants: A Memoir*. New York: Hyperion, 1996.

Power, Thomas G. *Play and Exploration in Children and Animals*. Mahwah, NJ: Lawrence Erlbaum Associates, Publishers, 2000.

Preston, Stephanie D., and Frans de Waal. "Empathy: Its Ultimate and Proximate Bases." *Behavioral and Brain Sciences* 25 (2002): 1–72.

Pribac, Teya Brooks. *Enter the Animal: Cross-Species Perspective on Grief and Spirituality*. Sydney: Sydney University Press, 2021.

Quinn, Rick. *Just Like Us: A Veterinarian's Visual Memoir of Our Vanishing Great Ape Relatives*. Seattle: Girl Friday Books, 2021.

Rachels, James. *Created from Animals: The Moral Implications of Darwinism*. New York: Oxford University Press, 1999.

Raihani, Nichola. *The Social Instinct: How Cooperation Shaped the Word*. New York: St. Martin's Press, 2021.

Recio, Belinda. *When Animals Rescue: Amazing True Stories about Heroic and Helpful Creatures*. New York: Skyhorse, 2021.

Regan, Tom. *The Case for Animal Rights*. Berkeley: University of California Press, 1983.

———. *Empty Cages: Facing the Challenge of Animal Rights*. New York: Rowman & Littlefield, 2005.

Riley, Sophie. *The Commodification of Farm Animals*. New York: Springer, 2022.

Rivera, Michelle. *Hospice Hounds*. New York: Lantern Books, 2001.

Rollin, Bernard E. *The Unheeded Cry: Animal Consciousness, Animal Pain, and Science*. New York: Oxford University Press, 1989.

Rosenberg, Gabriel N. *The 4-H Harvest: Sexuality and the State in Rural America*. Philadelphia: University of Pennsylvania Press, 2015.

Rothenberg, David. *Why Birds Sing: A Journey Through the Mystery of Birdsong*. New York: Basic Books, 2005.

Russell, W. M. S., and R. L. Burch. *The Principles of Humane Experimental Technique*. New York: Hyperion Books, 1959/1999.

Ryan, John, Patricia Vieira, and Monica Gagliano, eds. *The Mind of Plants: Narratives of Vegetal Intelligence*. Santa Fe, NM: Synergetic Press, 2021.

Safina, Carl. *Beyond Words: What Animals Think and Feel*. London: Picador, 2016.

Schoen, Allen. *Kindred Spirits: How the Remarkable Bond Between Humans & Animals Can Change the Way We Live*. New York: Broadway Books, 2001.

Scully, M. Dominion. *The Power of Man, the Suffering of Animals*. New York: St. Martin's Press, 2002.

Sebo, Jeff. *Saving Animals, Saving Ourselves: Why Animals Matter for Pandemics, Climate Change, and Other Catastrophes*. New York: Oxford University Press, 2022.

Shapiro, Kenneth J. *Animal Models of Human Psychology: Critique of Science, Ethics, and Policy*. Seattle: Hogrefe and Huber, 1998.

Sharpe, Lynne. *Creatures Like Us?* Exeter, UK: Imprint Academic, 2005.

Shepard, Paul. *Thinking Animals: Animals and the Development of Human Intelligence*. Athens: University of Georgia Press, 1998.

Singer, Peter. *Animal Liberation*. 2nd ed. New York: New York Review of Books, 1990.

———. *Animal Liberation Now: The Definitive Classic Renewed*. New York: HarperCollins, 2023.

Singer, Peter, and Jim Mason. *The Way We Eat: Why Our Food Choices Matter*. Emmaus, PA: Rodale, 2006.

Skutch, Alexander. *The Minds of Birds*. College Station: Texas A&M University Press, 1996.

Sneddon, Lynne U. "The Evidence for Pain in Fish: The Use of Morphine as an Analgesic." *Applied Animal Behaviour Science* 83 (2003): 153–62.

Sober, Elliott, and David S. Wilson. *Unto Others: The Evolution and Psychology of Unselfish Behavior*. Cambridge, MA: Harvard University Press, 1998.

Stallwood, Kim. *Preserving the History of the Animal Rights Movement*. Self-published, 2023. Https://kimstallwood.com/wp-content/uploads/2023/05/Report_Digital.pdf.

Stolkin, Gary. *Mykonos and Athena: A Furry Tale*. London: Austin Macauley, 2022.

Taylor, Maria. *Injustice: Hidden in Plain Sight the War on Australian Nature: Kangaroo, Koala, Emu…Hunted, Sold, Homeless…Where Lies Truce, Healing?* Australia: Maria Taylor, 2021.

Taylor, Paul W. *Respect for Nature: A Theory of Environmental Ethics*. Princeton, NJ: Princeton University Press, 1986/2011.

Tinbergen Niko. *Curious Naturalists*. Amherst: University of Massachusetts Press, 1984.

———. *The Herring Gull's World*. New York: Anchor Books, 1967.

———. *The Study of Instinct*. New York: Oxford University Press, 1951/1989.

Tobias, Michael, and Kate Solisti-Mattelon, eds. *Kinship with Animals*. Tulsa, OK: Council Oaks Books, 2006.

Turner, Jacky, and Joyce D'Silva, eds. *Animals, Ethics, and Trade*. London: EarthScan Publishing, 2006.

van Voorst, Roanne. *Once Upon a Time We Ate Animals: The Future of Food*. New York: HarperCollins, 2022.

von Frisch, Karl. *The Dance Language and Orientation of Bees*. Cambridge, MA: Harvard University Press, 1993.

Waldau, Paul, and Kimberley Patton, eds. *A Communion of Subjects: Animals in Religion, Science, and Ethics*. New York: Columbia University Press, 2006.

Waldman, Corey Lee. *Animals in Irish Society: Interspecies Oppression and Vegan Liberation in Britain's First Colony*. Albany, NY: SUNY Press, 2021.

Walton, Stuart. *A Natural History of Human Emotions*. New York: Grove Press, 2006.

Ward, Ashley. *The Social Lives of Animals*. New York: Basic Books, 2022.

Warwick, Clifford, Phillip C. Arena, and Gordon M. Burghardt, eds. *Health and Welfare of Captive Reptiles*. New York: Springer, 2023.

Warwick, Hugh. *Cull of the Wild: Killing in the Name of Conservation*. New York: Bloomsbury, 2024.

Webster, John. *Animal Welfare: Limping Towards Eden*. Oxford, UK: Blackwell Publishing, 2005.

Weil, Zoe. *Above All, Be Kind: Raising a Humane Child in Challenging Times*. British Columbia, Canada: New Society Publishers, 2003.

Wilson, David Sloan. *Darwin's Cathedral: Evolution, Religion, and the Nature of Society*. Chicago: University of Chicago Press, 2002.

Index

Page references in *italics* indicate illustrations or material contained in their captions.

abnormal repetitive behaviors (ARBs), 150
Achilles (octopus), 61
adaptation, 46, 114–15
admiration, 17
affection, 81
Agatha (goat), 32
aggression, 87
AI, 8
Albrecht, Glenn, 185
alligators, 4
Allyn, Rachel, 17
altered states, 72
altruism, 7, 96, 112
"Am I Blue?" (Walker), 60, 159
Amigo (elephant), 164
Ammann, Karl, 139
amphibians, 19, 29, 49
amygdala, 19, 46, 153
analogy, 48–50
Anderson, Roland, 61
Andrae, Margo, 158
anecdotes, 62, 124–25
Angell, Tony, 15

anger: as animal emotion, 55–56; field guide for observing, 87–90; as primary emotion, 18; as universal emotion, 17, 55
animal abuse: animal welfare and, 137, 142; beliefs behind, xiv, 13, 34–35; precautionary principle and, 136, 154, 185. *See also* animal suffering
animal consciousness, xi–xii, 3, 43, 194
animal emotions: anecdotal evidence for, 14–17; denial of, as "antiscientific," 3, 5, 39; "educated guesses" about, 3; evolutionary continuity and, xv–xvi, 5, 44–46; importance of, 28; interpreting, 57; paradigm shift needed in study of, 35–36; primary, 58; public interest in, 193; range of, 55–56; science of, 2, 13–14, 21, 57, 65; skepticism about, xvi, 21, 120–21; stereotypes about, 35; transparency of, 23–25, 31, 63, 189. *See also* animal emotions — observation of; animal sentience
Animal Emotions (blog; Bekoff), 1, 193
animal emotions — observation of: anger/aggression, 87–90; anthropomorphism and, 62; challenges involved in, 56–57; embarrassment,

animal emotions — observation of (*continued*)
86–87; eyes, 58, 60–61; facial expressions, 58, 59–60; field guide for, 55, 57–63; gratitude, 61–62; grief/sadness, 72–80; happiness/joy, 63–68; interpretation vs., 57; love, 80–86; mental impairments, 90–91; scent, 58–59; "showing off," 68–70

Animal Envy (Nader), 8

animal kill clock, 155

Animal Liberation (Singer), 34

Animal Models of Human Psychology (Shapiro), 144

animal research: captivity effects and, 148–51; environmental enrichment during, 150–51; fieldwork on wild animals, xiv, 146–48, 214; knowledge gained through, 143–44, 146, 151–52, 153–54; as "necessary," 145; noninvasive, 145, 151–53; numbers of animals used in, 141, 213; professional standards for, 145–46; suffering caused by, 134, 141–43, 144–45

animal rights, 4, 49, 166, 183–84

animals: action needed on behalf of, 185–89; as beneficial to humans, xv; captive, 50; as emotionless, 3, 5; "higher"/"lower" species, 34–35; individual, importance of, 11–12, 26, 34, 36–37, 131–32, 137, 177; justice for, 180–81; juvenile, as "cute," 126; lab, 21, 37; language used referring to, 12, 173–74, *174*; legal redefinitions of, 10, 37; morality in, 96, 97–98; naming, xiii, 2–3; personality differences in, 56–57; as "pests," 172–73; scientific objectification of, xiv, xv, 119; use of term, 12

Animals' Agenda, The (Bekoff and Pierce), 36, 133, 138, 139

animal sanctuaries, 32, 38

Animals Asia, 38, 67, 170

animal sentience: acknowledging, 4–5; animal welfare and, 5; conservation efforts and, 10–12, 26, 36; cross-species relationships and, 6–7; defined, 135; "degrees" of, 34–35; education about, 188; evolutionary continuity and, 5; knowledge translation gap regarding, 36–37, 133–34; legal declarations recognizing, xi–xii, 3–4, 9, 10; misconceptions about, 24–25; neuroanatomical structures facilitating, xii, 5–6; protections resulting from recognizing, 10–12; respecting, 8–9; scientific passion needed to support, 123–24; scientific research and, 35–37, 43–46, 147–48, 198; skepticism about, 117, 194; uncertainty about, 135–37; in unexpected species, 25–28

"Animal Sentience and Animal Welfare" (Webster), 135

animal studies, 10, 37

animal suffering: of captive reptiles, 49–50; of fish, 27; human-animal relationship complexity and, 33; in industrial agriculture, 63, 64, 134, 154–55, 156–58, 175; knowledge translation gap and, 133–34; sanitized language used to describe, 174–76; in scientific research, 134, 141–43, 144–45 (*see also* animal research); sentience "degrees" and, 5, 38; in sports/entertainment, 134, 160–61, 178; transparency of, 137; in wildlife conservation, 134, 171–73, 175, 214; in zoos, 134, 161–62, 163–65, 175

animal welfare, 196; animal well-being vs., 137–38; compassionate conservation and, 177; conferences about, 13; current standards of, 5, 10, 133, 185, 186, 188; Five Freedoms of, 138–40; industrial farm conditions and, 157–58; insect pollinators

and, 26; paradigm shift needed in, 137–41, 185–86, 188; as science, 138
animal well-being: animal welfare vs., 134, 137–38, 157–58; in captivity, 150, 154; caring about, 183–85; cognitive ethology and, 42; knowledge translation gap and, 36–37; in nature, 140; noninvasive research and, 154; play as essential to, 105; as science, 36, 138
Annie (mule), 84–85
Antarctica, 51
Anthes, Emily, 144
Anthropocene, 177, 188
anthropocentrism: as animal abuse cause, xiv, 13; animal morality and, 114; awareness of, 183–84, 185; compassionate conservation and, 177, 179; defined, 13; misconceptions resulting from, 24–25, 73, 114; problems of, 184; wildlife management programs and, 172. *See also* humanocentrism; speciesism
anthropomorphism, 89; alternatives to, 130; author's experience, 3, 211; benefits of, 130–32; biocentric, 130, 132, 211; cognitive ethology and, 118, 125–27; defined, 125–26; as evolved perceptual strategy, 21, 127; grief and, 73; scientific criticism of, 62, 120, 125, 127, 211; scientific double-talk about, 120–21, 128–30
"Anthropomorphism as a Conservation Tool" (Chan), 132
anthrozoology, 10, 198–99
antipredatory behavior, 100, 102
anxiety, 29
Aochan (snake), 6, 30–31
"ape exceptionalism," 69
apes, great, 68, 72, 112; scientific research on, xiv
apology, 106, 108
aquariums, 162, 167
Are We Smart Enough to Know How Smart Animals Are? (de Waal), 137

Arizona, 172
Ashe, Dan, 162
Association of Zoos and Aquariums, 129, 162, 166–67
Atkinson, Jenny, 79
Attenborough, David, 183–84
Australia, 4
Australian Pork Limited, 158
Austria, 4
autism, 90–91
awe, sense of, 56, 71–72

Babe (pig), 160
baboons, 6, 76–77, 89–90
Babyl (elephant), 6, 16, 21–22
Baez, Joan, 31
Baker-Carr, Janet, 73
Balcomb, Ken, 79
Balcombe, Jonathan, 64–65
Barnhill, Anet, 76
Baron-Cohen, Simon, 91
Barrett, Meghan, 26
Bartholomew, James, 156
Basel Zoo, 169
BBC News, 168
bear bile farms, 37–39
bears: captive, 33, 37–39, *39*, 168, 178; compassionate conservation and, 179; dancing, 178; empathy as shown by, 22, 38, *40*; humor as shown by, 67; hunting of, 10; play behavior of, 99–100; sanctuaries for, 38, *39*, *40*, 67; wildlife management programs targeting, 171
Becker, Marty, 29, 77
bees, 26, 64
behavioral ecology, 37
behavior patterns: emotions and, 5, 25; flexibility in, 43; framework for studying, 46–48; freedom to express, 138, 184; happiness/joy signaled by, 63; during social play, 102, 104, 106; stereotypies, 149–50; types of, 25

Bekoff, Marc, 39; as author, xi, xv–xvi, 36, 95, 123, 133, 138, 139, 167; blog of, 1, 193; as cognitive ethologist, 1–2, 41, 117–18
Belgium, 4
Bentham, Jeremy, 34, 38
Berry, Thomas, 185
Bexell, Sarah, 184
Beyond the North American Model of Wildlife Conservation (Heister), 171–72
Bianchi, Deb Pilley, 69
Bill (author's researcher friend), 120–21
biodiversity: compassionate conservation and, 177; industrial farms and loss of, 156; public interest in, 163; scientific research on, 218; of sentience, 26, 27–28, 148, 186
Biological Conservation, 131–32
bipolar animals, 91
Birch, Jonathan, 26–27
Bird Minds (Kaplan), 87
birds: anger/aggression as shown by, 87–88; animal welfare legislation and, 10, 37; author's studies of, 13; brain type of, 19; as companion animals, xii, 29; in cross-species relationships, 31; facial expressions of, 60; grief as shown by, 14–16, 48, 75–76; happiness as shown by, 64, 67–68; as lab animals, 141, 149; love as expressed by, 82; "love hormone" of, 81; play behavior of, 24, 65; relative brain size of, 24; self-awareness of, 24; as sentient animals, xii; speciesism and, 35; wild, fieldwork effects on, 214
Bitch: On the Female of the Species (Cooke), 57
Black Forest Animal Sanctuary, 84–85
Blackie (dog), 107
Black Lion (Mbatha), 8
Bochum (Germany), 121–22
boobies, 88
Boone (llama), 80

boredom, 149–50
Born Free Foundation, 128, 163
Boulder (CO) County Jail, 32–33
Bowlby, John, 48
brain: anger functions in, 87; captivity effects on, 149; laughter circuits in, 66; limbic system of, 19, 91; pleasure center in, 102; PTSD and, 73; relative size, 24; trust center in, 102
brain scans, 152
brain structure, 48–49
Braithwaite, Victoria, 27
Bridger (llama), 80
Brier-Patch Philosophy (Long), 125, 126
British Farm Animal Welfare Council, 138
Broken Shovels Farm Sanctuary, 32
Bronx Zoo, 166
Brookshire, Bethany, 173
Broom, Donald, 35, 138
Bryja, Gosia, 173, *174*
Bulgaria, 4
bullfighting, 9, 160
Burch, Rex, 145
Bureau of Land Management, 10
Burghardt, Gordon, 47, 49–50, 130
Burroughs, John, 134
Byosiere, Sarah-Elizabeth, 103

Cabanac, Michel, 49
cages: animal welfare and, 138, 142, 144; crush, 37–38, 39; environmental enrichment in, 150–51; repetitive stereotypies in, 149–50
California, 172
California Primate Research Center, 129
calves, 161
Cambridge Declaration on Consciousness (2012), xi–xii, 3
Cambridge University, 35
Campbell, Jeff, 7
Candy (elephant), 164, 169
canids, 82, 99, 102, 108. *See also* coyotes; dogs; foxes; wolves

"Can Insects Feel Pain?" (Gibbons), 26
Capitanio, John, 129
captivity, xii, 128, 163, 168
captivity effects, 148–51
Carbone, Larry, 141
Carew, Keggie, 183
Cassidy, Kira, 56–57
category E experiments, 141
Catherine Violet Hubbard Animal Sanctuary (Newtown, CT), 32
cats: as companion animals, xii, 29; as lab animals, 141; love as expressed by, 81; play behavior of, 99–100; in shelters, 166; "showing off" behavior of, 68, 70
cattle, 84–85, 155. *See also* cows; livestock
caudate nucleus, 102
causation, 46
cedar waxwings, 15
Center for Whale Research, 79
cetaceans, xii, 220. *See also* dolphins; orcas; whales
Chan, Alvin, 132
Charalambides, Eva, 159–60
Charlie (steer), 84–85
Chaser (dog), 69
chickens, 31, 63–64, 139, 156–58, 161
children, xiv, 31–33, 112, 162–63, 188, 195
Chile, 4
chimpanzees: anger/aggression as shown by, 90; anthropomorphism and, 130; awe as shown by, 71–72; captive, 130, 168, 169; embarrassment as shown by, 86; eyes of, 61; fieldwork with, 140; joy as expressed by, xiv, 63; as lab animals, 142; love as expressed by, 83; play behavior of, 99–100, 103; sanctuaries for, 72; scientific research on, xii–xiv, 3; self-awareness of, 24; as sentient animals, xiv, 24–25, 45–46, 55; "showing off" behavior of, 68–69
China: animal sanctuaries in, 38, 39, 67; animal welfare guidelines in, 138; bear bile farms in, 37–38, *39*
choice, freedom to express, 139
choughs, white-winged, 87
chronic wasting disease, 172
circuses, 138, 160
citizen scientists, 76
"clusterflocks," 43
cognitive dissonance, 173
cognitive ethology: analogy as used in, 48–50, 118; anecdotal evidence supporting, 118, 124–25; anthropomorphism and, 118, 125–27; author's experience, 1–2, 41, 117–18; defined, 42; fieldwork in, 41, 50–53, 122; legitimization of, 123–24; recent developments in, 1, 2–3, 42; research methods in, 46–48; scientific criticism of, 118–19, 120–22, 125, 147–48, 151; scientific research on, 13–14; uncertainty and, 122–23
cognitive skills, 105
Cole, Kathie, 29
Colorado Division of Wildlife, 172
Colorado River basin, 216
commitment, 81, 82–83
companion animals: agency and, 167; benefits/shortcomings of, 29–30; children and, 31–32, 195; numbers of, in US, 28–29; of scientists, 120–21; as sentient animals, xii, 8–9, 11, 21; stress experienced by, 11; trust of, 140–41. *See also* cats; dogs; *specific companion species*
comparative method, 44, 48, 87
compassion, 7, 30, 81, 135, 188–89. *See also* compassionate conservation; empathy
Compassionate Brain, The (Hüther), 25
compassionate conservation: anthropomorphism and, 131–32; author's experience, 177; defined, 177; development of, xii, 36–37, 177; education about, 181; guiding

compassionate conservation (*continued*)
principles of, 177, 179, 180–81; individual animals as important in, 11–12, 26, 36; paradigm shift in, 11, 36–37; reintroduction programs and, 179–80; scientific criticism of, 147, 177; wildlife management programs and, 171–72; zoo ethics and, 167

compassion fatigue, 183

composite signals, 51

conservation. *See* compassionate conservation; wildlife conservation

conservation behavior, 10

conservation biology, 37, 132

conservation psychology, 10

conservation washing, 163

contempt, 17

"Conversation with Frans de Waal, A" (de Waal), 117

Cooke, Lucy, 57, 169

"Cool Leaf! Study Records Chimp Showing Off Object in Human-Like Way" (*The Guardian*), 68–69

cooperation, 95, 96, 101–2, 107, 111–13

Copenhagen Zoo, 168

corvids, 14, 97

Covid-19 pandemic, 31

cows, 64, 81, 156–57, 161, 171. *See also* cattle; livestock

coyotes: author's studies of, 13, 51, 70, 99; autistic, 90–91; behavioral flexibility of, 43; grief as shown by, 165; love as expressed by, 82; play behavior of, 93, 94, 104, 107, 109; rescue centers for, 22–23; scientific research on, 109; "showing off" behavior of, 70; social bonds of, 164–65; wildlife management programs targeting, 172

crabs, 26–27

crayfish, 26–27

Created from Animals (Rachels), 44

Creatures Like Us? (Sharpe), 23

Croatia, 4

crocodiles, 23

cross-species relationships, 6–7

crows, xiv, 15, 65

Cruelty Free International, 141

crustaceans, 26–27

Current Biology, 101

Cyprus, 4

Czech Republic, 4

dairy industry, 159

Daisy to the Rescue (Campbell), 7

Dalai Lama, 111, 113

Damasio, Antonio, 17

"Dark Side of Zootopia, The" (Siebert), 161

Darwin, Charles: animal emotion studies of, 44; on cooperation and reproductive success, 93, 95, 107; emotions as identified by, 17, 18, 55–56, 63; evolutionary continuity theory of, 5, 44–46; scientific methods of, 44, 45; sexist bias of, 57. *See also* evolutionary continuity

Darwin's Cathedral (Wilson), 114

David Greybeard (chimpanzee), xii–xiii

Davis, Karen, 64, 175, 215

Davis, Neva, 70

Dawkins, Marian, 194, 211

deception, 103–4, 112–13

Dee (duckling), 31

deer, 171, 172, 176

Denmark, 4

Denver (CO), 32

Denver (CO) Women's Correctional Facility, 29

Denver Zoo, 163–64

depression, 26, 30, 150

Descartes' Error (Damasio), 17

Descent of Man and Selection in Relation to Sex, The (Darwin), 93

Detroit Zoo, 61, 166

devotion, 81

de Waal, Frans, 1, 3, 7, 117, 137

Diabetes Research Institute, 144

Index

Dickie, Leslie, 168
dignity, 10
dingoes, 82
discomfort, freedom from, 138, 139
disease, 172, 175; freedom from, 139
disgust, 17, 18, 55
Dodman, Nicholas, 21
Do Fish Feel Pain? (Braithwaite), 27
dogs: abandoned, PTSD experienced by, 73; author's studies of, 13, 50, 51–52, 102; awe as shown by, 72; behavioral flexibility of, 43; brain structure of, 46; as companion animals, xii, xiii, 29, 195; embarrassment as shown by, 86–87; empathy gap and, 11; facial expressions of, 59; grief as shown by, 74, 77; happiness as shown by, 20–21, 63, 65–67, 69; as lab animals, 141; love as expressed by, 81, 85–86; noninvasive research on, 152, 153; play behavior of, 20–21, 46–47, 57, 93, 94, 98–99, 102, 106–7, 110; rescue, 6; as sentient animals, xiv, 6, 7, 25, 46, 117; in shelters, 166; "showing off" behavior of, 68, 69–70; tool use of, 97
Dolan, R. J., 34
Dolly (elephant), 163–64, 169
dolphins: animal rights and, 166; captive, 168; cross-species relationships and, 30; happiness as shown by, 63; love as expressed by, 83; noninvasive research on, 152; scientific research on, xiv; as sentient animals, 3, 7, 24
domesticated animals, 9, 11–12, 64
Donkey (Tobias and Morrison), 41, 111
donkeys, 6, 41, 66, 73, 111
dopamine, 204
Douglas-Hamilton, Iain, 16, 52, 78
dualism, 34, 184
ducks, 31, 214
Dugatkin, Lee, 82
Dukore (gorilla), 98
Dundee Zoo, 169

eagles, 7, 88
ears, 25
ear tags, 214
Echo (elephant), 83–84
ecological interactions, 113
Ecuador, 4
egalitarianism, 110–11, 114–15
egrets, 88
Ehrlich, Paul, 135
Eisner, Gail, 215
Ekman, Paul, 63
Elephant Memories (Moss), 78
elephants: animal welfare concerns about, 196; anthropomorphizing of, 128–30; captive, 16, 128–29, 163–64, 166, 168, 196; empathy as shown by, 16–17, 21–22, 78; gender differences, 57; grief as shown by, 77–78, 164, 166; happiness as experienced by, 18; hunting of, 10; literary portrayals of, 125; love as expressed by, 83–84; sanctuaries for, 130, 166; scent as communication among, 58–59; scientific research on, xiv, 18, 52–53; self-awareness of, 24; as sentient animals, 6, 122; social bonds of, 164
Eliot (elephant), 84
elk, 172
Ely (elephant), 84
embarrassment, 17, 56, 86–87
emotional intelligence, 6, 7
Emotional Lives of Animals, The (Bekoff), xi, xv–xvi
emotions: communication with, 8; empathy and, 28; evolutionary continuity and, xv–xvi, 44–46, 95, 189; feelings vs., 17, 20; primary vs. secondary, 18–20, 21–22, 58; social, 17; as "social glue," 44–45; universal, 17–18, 55–56. *See also* animal emotions
empathy: as animal emotion, 7; anthropocentric views of, 25; companion animals and, 31–32; cross-species, 23, 30; elephant displays of, 16–17,

239

empathy (*continued*)
21–22; emotions as glue underlying, 28; evolutionary continuity and, 5; mammalian brains and, 6; morality and, 95, 96; play behavior and, 94; as universal emotion, 21–23
empathy fatigue, 183
empathy gap, 9–10, 11–12
endangered species: humor and, 68; rights of, 79; trophy hunting of, 10; zoo breeding programs and, 162
endocrinology, 87
Engh, Anne, 76
Enid (elephant), 84
entertainment, 134, 160–61, 178
environmental degradation, 155–56
environmental enrichment, 150–51
envy, 17
Estonia, 4
"ethical hunting," 175–76
ethics: animal sentience and, 14, 184; compassionate conservation and, 36–37, 177; defined, 134–35; hunter, 176; language use and, 173; morality vs., 95–96; objectivity vs., 138; vegan, 159–60; Western, as limited, 133; wildlife management and, 172; zoos and, 168
ethology, 87. *See also* cognitive ethology
European Association of Zoos and Aquaria, 168
Evil Bunny (rabbit), 31
evolutionary biology, 14
evolutionary continuity: animal emotions and, xv–xvi, 5, 44–46, 64–65, 73–74, 84, 189; empathy and, 5; morality and, 5, 94, 114–15; play and, 109–10; pleasure and, 64–65; "survival of the fittest" mentality and, 113
Expression of the Emotions in Man and Animals, The (Darwin), 44
extinction, freedom from, 139
Extravagance of Donkeys, An (Baker-Carr), 73

eye contact, 61
eyes, 25, 58, 60–61, 185
faces, 25, 58, 59–60
facilitation, 113
factory farms, 156–57. *See also* industrial farms
"fair chase," 176
fairness, 94, 95, 105–9
farm animals: animal welfare and, 138–40; anthropocentric portrayals of, 64; as sentient animals, xii. *See also* cattle; chickens; cows; domesticated animals; industrial farms; livestock; pigs; *specific farm animal*
fear: as animal emotion, xiv, 55–56, 215–16; freedom from, 138–39; human-induced, 147; observing, 61; as primary emotion, 18; as universal emotion, 17, 55
feelings: emotions vs., 17, 20; mammalian brains and, 5–6
Fehr, Ernst, 111
Ferd (dog), 99
Ferraro, Kristy, 177
fidelity, rule of, 140–41
Field, Kate, 146
Fifi (chimpanzee), 86
Figan (chimpanzee), 86
fight-or-flight response, 18
Finland, 4
Fiona (chimpanzee), 69
fish: animal welfare legislation and, 10, 37; author's studies of, 13; brain type of, 19; as companion animals, 29; facial expressions of, 60; farmed, numbers killed annually, 155; love as expressed by, 82; "love hormone" of, 81; self-awareness of, 24; as sentient animals, xii, 6, 27, 60, 216; speciesism and, 35; wild, dwindling numbers of, 156
Fisher, Helen, 82
fishing, 131, 156, 160

Index

Five (wolf), 60
Five Freedoms of Animal Welfare, 138–40
Flack, Jessica, 103
Flint (chimpanzee), xiv
Flo (chimpanzee), xiii, 86
Flow (dog), 55
fMRI studies, 82, 102, 152–53
Food and Drug Administration, 144
food industry, 142. *See also* industrial farms
forgiveness, 108, 114
Fourteen (wolf), 72–73
Fox, Michael W., 215
foxes, 25, 74–75, 82, 94
Fradkin, Philip, 156
France, 4
FRAPs (frenetic random activity periods), 20. *See also* zoomies, the
Fraser, David, 138
freedom, 138–40, 148–49, 167, 180–81
Freud (chimpanzee), 86
friendliness, 112–13
friends, 76–77, 81, 164
Frisch, Karl von, 46, 48
frontalin, 59
fun, 100–101, 139, 204
functional magnetic resonance imaging. *See* fMRI studies
funerals, 14–16
fur farms, 138
Furuviksparken zoo (Sweden), 169
future, thoughts about, 24–25

Garber, Paul, 112
gaze, 60–61
gender, 57
Germany, 4
Gibbons, Matilda, 26
Gifts of the Crow (Marzluff and Angell), 15
Gilbert, Daniel, 24
Gilka (chimpanzee), xiii
giraffes, 10, 139, 163, 168
Girona (Spain), 72

"Glass Walls" (McCartney), 154
Gluck, John, 37
glucocorticoids, 76–77
goats, 32
Gohan (hamster), 6, 30–31
Goliath (chimpanzee), xiii
Gombe National Park, xii–xiii, xiv
Goodall, Jane: animal-naming practice of, xiii, 3, 127; author's work with, 31, 33; chimpanzee research of, xii–xiv, 3; on chimpanzee sentience, 55, 66, 71–72, 86; chimpanzee tool use discovered by, 97; companion animals of, xii, xiii, 7; industrial farming critiques of, 215; relationship with chimpanzees, 140; Roots & Shoots program of, 31; scientific criticism of, xiii, 127
gorillas, 3, 8, 97–98
Gosling, Sam, 56
Grandin, Temple, 158
Grand Teton National Park, 51, 109, 164–65
gratitude, 61–62
Gray, Jenny, 168
Great Britain, 138
Greece, 4
Greek, Jean, 144
Greek, Ray, 55, 144
greenhouse gas emissions, 156
"greeting rumbles," 18
Gregg, Justin, 9
Gretel (goose), 48
grief: as animal emotion, xiv, 56, 83; avian displays of, 14–16; evolutionary continuity and, 73–74; maternal, 83; as universal emotion, 18, 73
Griffin, Donald R., 42
grooming, 112
grosbeaks, 13
Gross, Charles, 150–51
guilt, 17, 56
guinea pigs, xii, 141

Gulland, Frances, 62
gut reactions, 6, 61

habitat loss, 171
Hagel, Vincent, 15
hamsters, xii, 6, 30–31, 141
Hanif, Shahana, 166
Hansel (goose), 48
Hansen, Lawrence, 37
happiness: as animal emotion, 55–56; field guide for observing, 63–68; as primary emotion, 18; as universal emotion, 17, 18, 55. *See also* joy
Happy (elephant), 166
Hare, Brian, 112
Harlow, Harry, 143
Harrison, Bernard, 167–68
Harry (coyote), 90–91
Hauser, Marc, 86
Healing Power of Pets, The (Becker), 29
Health and Welfare of Captive Reptiles, The (ed. Warwick, Arena, and Burghardt), 49–50
heart, 13
Heart (dog), 6
Hebb, Donald, 130
Heister, Anja, 171–72
herons, 88
Herzog, Hal, 30
Hewa (elephant), 53
Hinde, Robert, xiii–xiv
hippocampus, 91
hippopotamuses, 85
honesty, 108–9
Hope (elephant), 164
Horowitz, Alexandra, 21, 107
horse racing, 160–61
horses, 6, 160–61
Hospice Hounds (Rivera), 29
How Animals Grieve (King), 73
"How Anthropomorphism Is Changing the Social Context of Modern Wildlife Conservation" (Manfredo et al.), 131–32

Hubbard, Catherine Violet, 32
Hubbard, Jenny, 32
human-animal relationships: animal sentience and, 8; "anthro-harmonic" approach to study of, 185; anthropomorphism and, 131; benefits of, 29–31; changing perspectives on, 10; children and, 31–33; companion animals and, 28–30, 31–32; compassionate conservation and, 177–79; complexity of, 9, 33–34; conflicts, 171, 177–79; consent within, 167; empathy gap in, 9–10, 11–12; ethics and, 133, 134–35; rules for, in nature, 140–41; scientific research on, 198–99; solastalgia, 185
"Human-Caused Mortality Triggers Pack Instability in Gray Wolves" (Cassidy et al.), 179
humane education, 188
Humane Society of the United States, 83, 128
humane washing, 157, 158, 176
human exceptionalism, 24–25
humanocentrism, 33–34, 64, 95, 97. *See also* anthropocentrism; speciesism
humans: animals as beneficial to, xv; brain structure of, 19, 102; emotional development in, 44–45; morality and, 97, 98; well-being of, and animal well-being, 184–85
humor, 24, 56, 66–68
Hungary, 4
hunger, freedom from, 138, 139
hunter ethics, 176
hunting, 9–10, 33, 131, 172, 175–76
Hutchins, Michael, 129
Hüther, Gerald, 25
hypothalamus, 46, 87

If Nietzsche Were a Narwhal (Gregg), 9
iguanas, 6, 49
India, 178–79, 196
Indiana Coyote Rescue Center, 22–23

indignation, 17
Indonesia, 165–66
industrial farms: animal suffering at, 9, 63–64, 134, 154–55, 156–58, 175; animal welfare improvements at, 157–58; dietary solutions to, 159–60; environmental impact of, 155–56, 216; exposés of, 154, 215; mass animal "depopulation" methods at, 175; meat consumption from, 160; precautionary principle and, 155
Infant Grief (Bowlby), 48
injury, freedom from, 139
insects: animal welfare legislation and, 10; as companion animals, 29; play behavior of, 64; as sentient animals, 25–26, 27
Institute for Animal Sentience and Protection, 132
Institutional Care and Use Committee meeting (Boston; 2006), 187–88
interconnection, 185
intuition, 6, 60
"invasive species," 175
invertebrates, xii, 29, 35, 83
Iran, 138
Ireland, 4
isotocin, 81
Italy, 4

jackals, 82
Jamieson, Dale, 167
Jane Goodall Institute, xii, 220
Jasim, Hasan, 15
Jasper (bear), 37–39, 39, 40, 67
Jaws (monkey), 145
jays, 13, 25
JB (chimpanzee), xiii
jealousy, 17, 19, 56, 152–53
Jerome (dog), 98–99
Jethro (dog), 52, 66–67, 69–70
Jingles (dog), 32
Joe (coyote), 91
Johnson, Lawrence, 140–41

JoJo (chimpanzee), 61
Jomeo (chimpanzee), xiii
Journal of the American Medical Association, 144
joy, 34, 46, 56, 63. *See also* happiness
justice: for animals, 180–81; animal sense of, 93–94, 95, 111; defined, 111; as natural, 114; restitutive, 140, 141
Justice for Animals (Nussbaum), 139

Kalandars, 178
Kaplan, Gisela, 24, 87, 147
Kentucky Derby, 160–61
Kerala (India), 196
keystone species, 2
"killing for fun" wildlife contests, 172
Kindred Spirits (Schoen), 29
King, Barbara, 73
knowledge translation gap, 36–37, 133–34, 154, 170
Knoxville Zoo, 128, 129
Kobuk (dog), 85–86
Koko (gorilla), 8
Kristal (chimpanzee), xiv
Kristof, Nicholas, 157

lab animals, 33, 37, 213
laboratories, 50
Lahvis, Garet, 37
Lambert, CeAnn, 22–23
language: importance attached to, 7; rewilding of, 174; sanitized, 174–76; terminology used referring to animals, 12, 173–74, *174*; third-person, in scientific research, 119; trust and, 113
"Language and the Knife" (Bryja), 173
Larry (lamb), 31
Latvia, 4
laughter, 65–66
Lawton, Graham, 10
Leakey, Louis S. B., xiii
leeches, 83
lemurs, 112

Leonardo da Vinci, 59
leopards, 10, 178–79
Levinson, Boris, 32
limbic system, 19, 63, 91
lions, 10, 30, 163, 168–69
Lithuania, 4
livestock, 155, 156, 157, 171, 175, 216. *See also* cattle; cows
llamas, 79–80
lobsters, 26–27, 216
Lolita (dog), 70
London, Jack, 134
London Zoo, 45–46
loneliness, 30
Long, William J., 125, 126, 134
longing, 19
Lorenz, Konrad, 21, 46, 47, 48, 126
Lori (dog), 86–87
Los Angeles Zoo, 128, 129, 130, 163, 166
love, 17; as animal emotion, 56; cross-species, 84–85; evolutionary continuity and, 84; field guide for observing, 58–59, 80–86; filial, 81; human understanding of, 80–81; maternal, 81, 83–84; romantic, 81–83; as universal emotion, 73
"love hormone," 46, 81. *See also* oxytocin
loyalty, 81
Lucy (wolf), 91
Lupey (wolf), 104
Luxembourg, 4

Mac (macaw), 67–68
Macein, David Nieto, 88
MacLean, Paul, 19
Maddy (dog), 20, 69–70
magpies, 14–15, 24
Malta, 4
mammals: brain types possessed by, 19; captive, 163; facial expressions of, 59; limbic system of, 63; love as expressed by, 82–83; "love hormone" of, 81; neuroanatomical structures of, 5–6; play behavior of, 66; as sentient animals, xii

Manfredo, Michael, 131–32
Man Meets Dog (Lorenz), 21
manta rays, 24
Marco (chimpanzee), 72
Marcus, Eric, 215
Marine Mammal Center, 62
Marius (giraffe), 168
marmosets, 82, 150–51
Maryland, 172
Marzluff, John, 15
Masai Mara National Reserve (Kenya), 89
Mason, Georgia, 150
Mason, Jim, 159, 215
Massachusetts, 172
Matheson, David, 101
mating behavior, 81–83, 100, 102
Matsusaka, Takahisa, 66
Maxine (dog), 72
Maxine (elephant), 166
May, Brian, 13, 128
Mbatha, Sicelo, 8
McCartney, Paul, 154
McComb, Karen, 78
meat diet, 154, 156, 160
meatpacking industry, 154, 157
Meat Paradox, The (Percival), 160
media coverage, 79
"Meet Five, Nine, and Fourteen, Yellowstone's Heroine Wolves" (Smith), 72–73
Memoirs of Childhood and Youth (Schweitzer), 133
Menigoz, Mick, 62
mental impairments, 90–91
mesotocin, 81
methane, 156
mice: animal welfare legislation and, 10, 37, 142, 185; empathy as shown by, 22–23; as lab animals, 21, 141, 142–43, 144, 147, 149, 185; noninvasive research on, 152; as sentient animals, 6, 142–43; speciesism and, 35; wild, fieldwork effects on, 214
Mickey (dog), 20

Midgley, Mary, 129–30
Milo, Ron, 155
Mimi (elephant), 164, 169
mind, theory of, 68, 110
mink, 138
Moby Dick (Melville), 125
mollusks, 26–27
Mom (coyote), 164–65
Mombasa (Kenya), 85
Mona Foundation, 72
Mona Lisa (Leonardo da Vinci), 59
monkeys: behavioral flexibility of, 43; cooperative activities of, 112; embarrassment as shown by, 86; empathy as shown by, 22; facial expressions of, 59; fear as shown by, xiv; as lab animals, 141, 143, 144–45, 149; noninvasive research on, 152; scientific research on, xiv
monogamy, 82–83
Moon Bear Rescue Centre, 38, 39, 67
morality: as animal behavior, 7, 93–95; anthropocentric view of, 114; basic ingredients of, 95; ethics vs., 95–96; evolutionary continuity and, 5, 94, 114–15; play behavior and, 96, 97–98, 111, 114, 115; social, 42. *See also* cooperation; fairness; justice
Morrison, Jane Gray, 41, 67, 111
Moskito, James, 62
Moss, Cynthia, 78, 83–84, 122
mountain lions, 61, 74
Ms. Lopsided (chicken), 31
mules, 84–85
murder, 175
music therapy, 166
musth, 58–59
Mutsugoro Okoku Zoo (Tokyo, Japan), 6, 30–31
mutualism, 131, 132, 175
Mzee (tortoise), 85

Nader, Ralph, 8
Nakuru, Lake, 71
National Institutes of Health (NIH), 150, 151
National Science Foundation, 122
Natural History of Human Emotions, A (Walton), 17
Nature, xiii
nature: animal well-being in, 140; human-animal relationships in, 140–41; human relationship with, 184–85, 187, 188
neocortical brain, 19
Netherlands, 4, 165–66
neurobiology, 14, 87
neurochemicals, 87
Newberry, Ruth, 105
New Mexico, 172
New Scientist, 101
New York Supreme Court, 166
New York Times, 157, 166
New York Times Magazine, 60–61
New Zealand, 4, 30
Ngwino (gorilla), 98
Nick (baboon), 89–90, 127
Nobel Prizes, 46
nociceptors, 26
Nonhuman Rights Project, 166
noninterference, rule of, 140
nonmaleficence, rule of, 140
norepinephrine, 204
"nuisance" animals, 172–73, 175, 178–79
Nussbaum, Martha C., 139, 180
nuzzling, 62

Obi (cat), 70
objectivity, xiii, 138
ocean acidification, 156
octopuses, xii, 9, 26–27, 61
"odd couples," 6–7, 30–31, 194
Olly (chimpanzee), xiii
One Health initiatives, 184
One Welfare initiatives, 184
On Talking Terms with Dogs (Rugaas), 101
ontogeny, 46

opioids, 101
optimism, 26
orangutans, 45, 97, 165–66, 168, 169
orcas, 79, 83, 88–89
Oregon National Primate Research Center, 145
Oregon Regional Primate Research Center, 144–45
Origins of Virtue, The (Ridley), 114
oryx, 30
Oscar (dog), 74
ostracism, 109
otters, 6
Owen (hippo), 85
Owens, Delia, 60
oxytocin, 46, 81

Pablo (chimpanzee), 37
pain: animal sentience and, 26, 27, 216; freedom from, 139; of individual animals, 34; intentional, in animal research, 33, 134, 143, 154; maternal, 83; plants and, 28. *See also* animal suffering
paleomammalian brain, 19
Panama, 49
pandas, 169
Panksepp, Jaak, 3, 101
parrots, xiv, 67–68, 149
Passion for Justice, A (Solomon), 114
Patty (elephant), 166
pelicans, 88
Pellis, Sergio, 105
penguins, 13, 51
Peninsula Valdis (Argentina), 82
Pepsi (dog), 77
Percival, Rob, 160
Performing Animal Welfare Society, 130
pessimism, 26
"pests," 172–73, 175
Pests (Brookshire), 173
Peterson, Dale, 139
pets. *See* companion animals
PET scans, 152

pet therapy, 32
pheromones, 58
Philadelphia Zoo, 166
Phoenix (orca), 79
Pierce, Jessica, 36, 95, 96, 133, 138, 139, 167, 187
pigeon shoots, 160
pigs: captive, behavior of, 149; as farm animals, 143, 156–57, 158, 185; pain experienced by, 143; play behavior of, 103; scientific research on, xiv, 143; as sentient animals, xiv, 81; slaughtering of, 155, 185
plants, 28
play bows: abnormal behavior with, 90; as apologies, 96, 108; defined, 47; as play initiation signals, 47, 102, 104, 108; as "reset" signals, 103; scientific research on, 47–48, 57
play inhibition, 104–5
play panting, 66
play / play behavior: author's studies of, 93, 94, 99, 102–3; benefits of, 105–6, 204; breakdown of, 106–7, 108–9; cooperation and, 111–13; dynamics of, 102–5, 107–9; as egalitarian, 110–11; evolutionary continuity and, 109–10; fairness during, 96–97, 105–9; "Five Ss of," 100; framework for studying, 46–48; "language" of, 98–99; morality and, 96, 97–98, 111, 114, 115; reasons for, 100–102; rough-and-tumble, 88, 110; scientific research on, 106, 204; "Six Fs of," 100; vocalizations during, 66. *See also* play bows
play signals, 103–5. *See also* play bows
pleasure, 64–65
Poland, 4
Pollack, Henry, 135–37
Pollan, Michael, 215
polo, 160
Poole, Joyce, 18
pork industry, 157, 158

Portugal, 4
"Postzoo Future, A" (Bekoff and Pierce), 167
poultry industry, 157
prairie dogs, 75, 76
precautionary principle, 135–37, 154, 155, 185
predatory behavior, 100, 102
pride, 17, 68
primates: brain structure of, 19; human-like behavior of, 76; as lab animals, 149; moral behavior of, 97–98, 106; noninvasive research on, 152; play behavior of, 94; scientific research on, 94, 106; "showing off" behavior of, 68, 69. *See also* chimpanzees; monkeys; *specific species*
Primate's Memoir, A (Sapolsky), 89
Principles of Humane Experimentation Technique, The (Russell and Burch), 145
privacy, freedom to have, 139
"problem locations," 178–79
pronoun usage, xiii, 12, 173–74
prosimians, diurnal, 112
Provine, Robert, 66
Psychology Today, 1, 193
PTSD, 73, 91
"Publication Reform to Safeguard Wildlife from Researcher Harm" (Field), 146
Puerto Rico, 163
Pumpernickel (llama), 80
punishment, 106

quail, 75–76
Question of Animal Awareness, The (Griffin), 42

rabbits, 31, 141, 150
Rachels, James, 44
radio transmitters, 214
rats: animal welfare legislation and, 10, 37, 142, 185; behavioral flexibility of, 43; happiness as shown by, 66; as lab animals, 141, 149, 185; play behavior of, 99–100; scientific research on, xiv; speciesism and, 35; suicidal, 21, 56; tool use of, 97
rattlesnake roundups, 160
rattlesnakes, 49
ravens, 15
Reader's Digest, 62
Recio, Belinda, 7
reciprocity, 94, 96, 111
regret, 19
rehabilitation centers, 178
Reijinder, Lucas, 156
reintroduction projects, 163
Reis, Marlon, 132
religious behavior, 71
Reno (dog), 120–21
reptiles: brain type of, 19; captive, 49–50; as companion animals, 29; facial expressions of, 60; "love hormone" of, 81; sensory pleasure maximized by, 49
reptilian brain, 19
Respect for Nature (Taylor), 140
restitutive justice, rule of, 140, 141
Reuben (baboon), 89–90
revenge, 87
"Review of the Evidence of Sentience in Cephalopod Molluscs and Decapod Crustaceans" (Birch et al.), 26–27
rewilding efforts, 10, 186, 187
rhinoceroses, 10, 71, 163
Ridley, Matt, 114
"Rights of Nature" movement, 10
Rilling, James, 102, 152
Rivera, Michelle, 29
Robinson, Jill, 38, 67, 170
rodents, 29
rodeos, 160, 161
role-reversing, 105, 108
Romania, 4
Roosevelt, Theodore, 134
Roots & Shoots program, 31

Rose, Naomi, 83
Rosie (elephant), 164
Rossell, Matt, 144–45
Roxy (dog), 107
Ruby (elephant), 128–29, 130, 166
Rufus (dog), 86–87
Rugaas, Turid, 101
Russell, William, 145
Rusty (dog), xii, xiii, 7
"ruth," 187
Rwema (gorilla), 98

Sadie (dog), 74
"Sad Lot of Lab Chimps, The" (Goodall and Greek), 55
sadness: as animal emotion, 55–56; as primary emotion, 18; as universal emotion, 17, 55. See also grief
"safe zones," 139
Samburu Reserve (Kenya), 30, 52–53, 139
Sandy Hook shooting (2012), 32
San Francisco Zoo, 166
Sanjay Gandhi National Park, 178–79
San Juan Island (WA), 79
Santa Anita Racetrack, 161
Sapolsky, Robert, 89–90, 127
Satyanarayan, Kartick, 178
Saudi Arabia, 90
Save Me Trust, 128
scent, 25, 58–59
Schaller, George, 94
Schoen, Allen, 29
Schusterman, Ron, 90
Schweitzer, Albert, 133
Science, 34
science: animal sentience and, 198; animal suffering in, 134; ethics and, 36–37, 134–35; "hard" vs. "soft," 151; paradigm shift needed in, 35–37; sentimentalization of, 134; third-person language used in, 119; uncertainty in, 135; values attached to, xiii, 118–19. See also animal research

"science sense," 14
Scully, Matthew, 215
sea lions, 83
seals, 7
sea sanctuaries, 167
Secret Life of Cows, The (Young), 63
Secret Social Lives of Reptiles, The (Doody, Dinets, and Burghardt), 49
Secrets of the Savannah (Owens), 60
self-awareness, 24, 51–52
self-handicapping, 104–5, 108
sentience: biodiversity of, 26, 27–28, 148, 186; defined, 4; as "keystone concept," 2. See also animal sentience
serotonin, 87, 204
Seshamani, Geeta, 178
Seton, Ernest Thompson, 134
sexism, 57
shame, 17
Shapiro, Kenneth, 144
Sharpe, Lynne, 23
sheep, 31, 81, 171
shelters, 166
Shepard, Paul, 132
"showing off," 56, 68–70
shrimp, 26–27
siblicide, 88
sibling conflicts, 87–88
Siebert, Charles, 60–61, 161
Sierra (baboon), 76
Simonet, Patricia, 65–66
Singapore Zoo, 167
Singer, Peter, 34, 159, 215
slaughterhouses, 156, 157, 185. See also industrial farms
Slobodchikoff, Con, 75–76
Slovakia, 4
Slovenia, 4
Smile of a Dolphin, The (Bekoff), 123
Smith, Doug, 60, 72–73
Smithsonian Institution, 122
snakes, 6, 30–31
social bonds, 76–77, 84–85
social organization, 6

Index

Socrates, 134–35
solastalgia, 185
Solomon, Robert, 114
Soret, Sam, 156
Soul (dog), 6
Spain, 4, 9–10, 88–89
speciesism, 114, 173, 175. *See also* anthropocentrism; humanocentrism
Species Survival Plan, 162
Sperm Whales (Whitehead), 125
spiders, 29
spindle cells, 6
Spinka, Marek, 105
sports, 134, 160–61
squid, 26–27
stereotypies, 128, 149–50
stress: animal research and, 50, 122, 146, 148–51; in baboons, 76–77; brain structure and, 73; captivity-induced, 148–51; of companion animals, 11; deaths from, at industrial farms, 157; human-animal relationships and, 29–30; plants and, 28
Stumbling on Happiness (Gilbert), 24
subjectivity, 118–19, 173–74
Suma (elephant), 166
surprise, 17, 18, 55
survival, 5
"survival of the fittest," 94–95, 113
Survival of the Friendliest (Hare and Woods), 112
Sussman, Robert, 112
Sutherland (chimpanzee), 69
Sweden, 4, 168–69
Switzerland, 4
Swope, Rick, 61
Sylvia (baboon), 76
sympathy, 17, 30

Taffy (llama), 80
Tahlequah (orca), 79, 83
tails, 23, 25
Taylor, Paul, 140
tenderness, 81

Teresa (elephant), 78
testosterone, 87
Thinking Animals (Shepard), 132
Think Like a Vegan (Charalambides), 159–60
thirst, freedom from, 138, 139
Thirteen (wolf), 72
three-brain theory, 19
Three Rs, the, 145–46
tickles, 66
tigers, 163, 168
Tika (dog), 85–86
Time magazine, 163
Tina (elephant), 78
Tinbergen, Niko, 46, 47, 48, 91, 94
Tobias, Michael, 41, 67–68, 111
Tolstoy, Leo, 37
Tomasello, Michael, 112
tools, use of, 97
tortoises, xii, 85
toxoplasmosis, 56–57
translocation programs, 178–80
trapping, 131, 171
Travers, Bill, 128
Trista (elephant), 78
trophy hunting, 9–10
trust, 95, 100, 140; breaking of, 106–7, 108–9
Tulullah (elephant), 78
turkeys, 156–57
turtles, 49

UCLA, 29
Uncertain Science . . . Uncertain World (Pollack), 135–37
uncertainty, 119, 135–37
unfairness, 114
United Kingdom, 4, 27, 163
United Nations Harmony with Nature Knowledge Network, 177
United Poultry Concerns, 64, 175
United States: adverse drug reaction deaths in, 144; animal protections in, 49; "animals" as defined in, 10,

United States (*continued*) 37, 142, 185; animal sentience legally unrecognized in, 4; animal welfare legislation in, 10 (*see also* United States Federal Animal Welfare Act); food animals slaughtered in, 155, 185; gun violence in, 32; "killing for fun" wildlife contests in, 172; lab animals in, 141; pet industry in, 28–29; western, industrial agriculture and water supply in, 156

United States Agriculture Department, 155

United States Federal Animal Welfare Act: "animals" as defined in, 10, 27, 37, 142, 185; animal suffering unrelieved by, 142–43, 188; enforcement of, 187–88

United States Fish and Wildlife Service, 10

United States Justice Department, 143

United States Supreme Court, 158

University of Amsterdam, 59

University of Bristol, 65

University of California–Davis, 129

University of California–Santa Cruz, 90

University of Denver, 132, 184

University of Miami Miller School of Medicine, 144

University of Pennsylvania, 76

University of Wisconsin, 143

Up-ears (dog), 107

Vancouver Island (Canada), 83

veal industry, 159

vegan diet, 159, 174

vegan ethics, 159–60

vegetarianism, 159, 174

"ventilation shutdown," 175

ventromedial prefrontal cortex, 73

Vermont, 172

vocalizations, 65–66

Vucetich, John, 184

Walker, Alice, 60, 159

wallabies, 104–5

Walton, Stuart, 17

Ward, Ashley, 43

Warneken, Felix, 112

Washington State, 171, 172, 175

waterfall dances, 71–72

Watson, Duncan, 105

Way We Eat, The (Singer and Mason), 159

Webb, Betsy, 79–80

Webster, John, 65, 135

Wedajo, Wondimu, 30

Weil, Zoe, 181, 188

Wemelsfelder, Françoise, 149–50

Whale Museum, 79

whales: anger/aggression as shown by, 88–89; captive, 168; gratitude as expressed by, 61–62; grief as shown by, 79; literary portrayals of, 125; love as expressed by, 82, 83; scientific research on, xiv; as sentient animals, 3, 6; spindle cells possessed by, 6. *See also* orcas

When Animals Rescue (Recio), 7

Where Have All the Animals Gone? (Peterson), 139

White as the Waves (Baird), 125

White Bone, The (Gowdy), 125

Whitehead, Hal, 125, 126

Why We Love (Fisher), 82

wilderness, 170, 171

Wild Justice (Bekoff and Pierce), 95

wildlife conservation: animal suffering in, 134, 171–73, 175, 214; anthropomorphism and, 131, 132; ethics of, 177, 179; human-animal relationships and, 10; public interest in, 163; purported purpose of, 177; sanitized language used to describe, 175; wildlife corridors, 171; zoo programs, 162–63, 169. *See also* compassionate conservation

wildlife corridors, 171
wildlife-killing contests, 172
wildlife parks, 163
Wildlife Services, 171
Wildlife SOS India, 178
Wilson, David Sloan, 114
wing tags, 214
Wittemyer, George, 52, 53
wolves: author's studies of, 13, 99; behavioral flexibility of, 43; bipolar, 91; captive, 104, 168, 169; eyes of, 60; gender differences, 57; grief as shown by, 72–73; love as expressed by, 82; personality differences in, 56–57; play behavior of, 57, 93, 94, 104, 107; as predators, 172; reintroduction programs, 179–80; scientific research on, xiv, 56–57; as sentient animals, 25; social bonds of, 179–80; wildlife management programs targeting, 171, 172, 175
Woods, Vanessa, 112
World Bird Sanctuary (Valley Park, MO), 7
World Society for the Protection of Animals conference (Rio de Janeiro; 2006), 27
wrasses, 24
Würsig, Bernd, 82
Wyler, Gretchen, 128–29

Yamamoto, Kazuya, 31
Yellowstone National Park, 56
Yellowstone Wolf Project, 56, 60, 72–73
Young, Rosamund, 63
"Your Dog Feels as Guilty as She Looks" (de Waal), 1, 3
Y2Y conservation initiative, 171

Zagreb Zoo, 166
Zarina, Leva, 103
Zeke (dog), 20, 69–70
Zoe (dog), 69–70
zoochosis, 128
Zoo Ethics (Gray), 168
zoomies, the, 20, 69, 70
zoos: accreditation/reaccreditation of, 167; animal relocations among, 163–64, 169–70; animal suffering at, 134, 161–62, 163–65; breeding programs at, 162, 163, 168, 169; conservation programs at, 162–63; as educational, 162; elephant exhibits phased out at, 166; emotional enrichment programs at, 165–66; knowledge translation gap and, 170; "naturalistic enclosures" at, 139; reintroduction projects at, 163; rewilding, 167; suggested reforms for, 166–70; "surplus animals" killed at, 163, 168–69, 175
Zoos Victoria, 168
zoothanasia, 163, 168–69, 175

About the Author

Marc Bekoff is Professor Emeritus of Ecology and Evolutionary Biology at the University of Colorado, Boulder. He has published thirty-one books (or forty-one if you count multivolume encyclopedias) and won many awards for his research on animal behavior, animal emotions (cognitive ethology), compassionate conservation, and animal protection. He has worked closely with Jane Goodall, is cochair of the ethics committee of the Jane Goodall Institute, and is a former Guggenheim Fellow. He also works closely with inmates at the Boulder County Jail. In June 2022, Marc was recognized as a "Hero" by the Academy of Dog Trainers. His latest books include *Dogs Demystified: An A-to-Z Guide to All Things Canine*, *Canine Confidential: Why Dogs Do What They Do*, and, with Jessica Pierce, *The Animals' Agenda: Freedom, Compassion, and Coexistence in the Human Age*, *Unleashing Your Dog: A Field Guide to Giving Your Canine Companion the Best Possible Life*, and *A Dog's World: Imagining the Lives of Dogs in a World without Humans*. Marc also publishes regularly for *Psychology Today*. In 1986, Marc won the Master's Tour du Haut, aka the age-graded Tour de France. His homepage is marcbekoff.com.

NEW WORLD LIBRARY is dedicated to publishing books and other media that inspire and challenge us to improve the quality of our lives and the world.

We are a socially and environmentally aware company. We recognize that we have an ethical responsibility to our readers, our authors, our staff members, and our planet.

We serve our readers by creating the finest publications possible on personal growth, creativity, spirituality, wellness, and other areas of emerging importance. We serve our authors by working with them to produce and promote quality books that reach a wide audience. We serve New World Library employees with generous benefits, significant profit sharing, and constant encouragement to pursue their most expansive dreams.

Whenever possible, we print our books with soy-based ink on 100 percent postconsumer-waste recycled paper. We power our Northern California office with solar energy, and we respectfully acknowledge that it is located on the ancestral lands of the Coast Miwok Indians. We also contribute to nonprofit organizations working to make the world a better place for us all.

Our products are available wherever books are sold.

customerservice@NewWorldLibrary.com
Phone: 415-884-2100 or 800-972-6657
Orders: Ext. 110
Fax: 415-884-2199
NewWorldLibrary.com

Scan below to access our newsletter
and learn more about our books and authors.